3-D Geometry

Seeing Solids and Silhouettes

Grade 4

Also appropriate for Grade 5

Michael T. Battista
Douglas H. Clements

Developed at TERC, Cambridge, Massachusetts

Dale Seymour Publications®
White Plains, New York

The *Investigations* curriculum was developed at TERC (formerly Technical Education Research Centers) in collaboration with Kent State University and the State University of New York at Buffalo. The work was supported in part by National Science Foundation Grant No. ESI-9050210. TERC is a nonprofit company working to improve mathematics and science education. TERC is located at 2067 Massachusetts Avenue, Cambridge, MA 02140.

This project was supported, in part, by the

National Science Foundation

Opinions expressed are those of the authors and not necessarily those of the Foundation

Managing Editor: Catherine Anderson

Series Editor: Beverly Cory

Revision Team: Laura Marshall Alavosus, Ellen Harding, Patty Green Holubar, Suzanne Knott, Beverly Hersh Lozoff

ESL Consultant: Nancy Sokol Green

Production/Manufacturing Director: Janet Yearian

Production/Manufacturing Coordinator: Joe Conte

Design Manager: Jeff Kelly

Design: Don Taka

Illustrations: DJ Simison, Carl Yoshihara

Cover: Bay Graphics

Composition: Archetype Book Composition

This book is published by Dale Seymour Publications®, an imprint of Addison Wesley Longman, Inc.

Dale Seymour Publications
10 Bank Street
White Plains, NY 10602
Customer Service: 1-800-872-1100

DALE SEYMOUR PUBLICATIONS®

Printed on Recycled Paper

Order number DS43892
ISBN 1-57232-745-6

9 10-ML-02

T E R C

Principal Investigator   Susan Jo Russell

Co-Principal Investigator   Cornelia Tierney

Director of Research and Evaluation   Jan Mokros

Curriculum Development
Joan Akers
Michael T. Battista
Mary Berle-Carman
Douglas H. Clements
Karen Economopoulos
Ricardo Nemirovsky
Andee Rubin
Susan Jo Russell
Cornelia Tierney
Amy Shulman Weinberg

Evaluation and Assessment
Mary Berle-Carman
Abouali Farmanfarmaian
Jan Mokros
Mark Ogonowski
Amy Shulman Weinberg
Tracey Wright
Lisa Yaffee

Teacher Support
Rebecca B. Corwin
Karen Economopoulos
Tracey Wright
Lisa Yaffee

Technology Development
Michael T. Battista
Douglas H. Clements
Julie Sarama
Andee Rubin

Video Production
David A. Smith

Administration and Production
Amy Catlin
Amy Taber

*Cooperating Classrooms
for This Unit*
Kerri Strobelt
Joyce Ungar
*Hudson Local School District
Shaker Heights, OH*

Brenda Spivey
*Shaker Heights School District
Shaker Heights, OH*

Michele de Silva
*Boston Public Schools
Boston, MA*

Kathleen D. O'Connell
*Arlington Public Schools
Arlington, MA*

Consultants and Advisors
Elizabeth Badger
Deborah Lowenberg Ball
Marilyn Burns
Ann Grady
Joanne M. Gurry
James J. Kaput
Steven Leinwand
Mary M. Lindquist
David S. Moore
John Olive
Leslie P. Steffe
Peter Sullivan
Grayson Wheatley
Virginia Woolley
Anne Zarinnia

Graduate Assistants
Joanne Caniglia
Pam DeLong
Carol King
Kent State University

Rosa Gonzalez
Sue McMillen
Julie Sarama
Sudha Swaminathan
State University of New York at Buffalo

Revisions and Home Materials
Cathy Miles Grant
Marlene Kliman
Margaret McGaffigan
Megan Murray
Kim O'Neil
Andee Rubin
Susan Jo Russell
Lisa Seyferth
Myriam Steinback
Judy Storeygard
Anna Suarez
Cornelia Tierney
Carol Walker
Tracey Wright

CONTENTS

TEACHER NOTES

WHERE TO START

The first-time user of *Seeing Solids and Silhouettes* should read the following:

When you next teach this same unit, you can begin to read more of the
background. Each time you present the unit, you will learn more about
how your students understand the mathematical ideas.

Investigations in Number, Data, and Space® is a K–5 mathematics curriculum with four major goals:

- to offer students meaningful mathematical problems
- to emphasize depth in mathematical thinking rather than superficial exposure to a series of fragmented topics
- to communicate mathematics content and pedagogy to teachers
- to substantially expand the pool of mathematically literate students

The *Investigations* curriculum embodies a new approach based on years of research about how children learn mathematics. Each grade level consists of a set of separate units, each offering 2–8 weeks of work. These units of study are presented through investigations that involve students in the exploration of major mathematical ideas.

Approaching the mathematics content through investigations helps students develop flexibility and confidence in approaching problems, fluency in using mathematical skills and tools to solve problems, and proficiency in evaluating their solutions. Students also build a repertoire of ways to communicate about their mathematical thinking, while their enjoyment and appreciation of mathematics grows.

The investigations are carefully designed to invite all students into mathematics—girls and boys, members of diverse cultural, ethnic, and language groups, and students with different strengths and interests. Problem contexts often call on students to share experiences from their family, culture, or community. The curriculum eliminates barriers—such as work in isolation from peers, or emphasis on speed and memorization—that exclude some students from participating successfully in mathematics. The following aspects of the curriculum ensure that all students are included in significant mathematics learning:

- Students spend time exploring problems in depth.
- They find more than one solution to many of the problems they work on.

- They invent their own strategies and approaches, rather than rely on memorized procedures.
- They choose from a variety of concrete materials and appropriate technology, including calculators, as a natural part of their everyday mathematical work.
- They express their mathematical thinking through drawing, writing, and talking.
- They work in a variety of groupings—as a whole class, individually, in pairs, and in small groups.
- They move around the classroom as they explore the mathematics in their environment and talk with their peers.

While reading and other language activities are typically given a great deal of time and emphasis in elementary classrooms, mathematics often does not get the time it needs. If students are to experience mathematics in depth, they must have enough time to become engaged in real mathematical problems. We believe that a minimum of 5 hours of mathematics classroom time a week—about an hour a day—is critical at the elementary level. The scope and pacing of the *Investigations* curriculum are based on that belief.

We explain more about the pedagogy and principles that underlie these investigations in Teacher Notes throughout the units. For correlations of the curriculum to the NCTM Standards and further help in using this research-based program for teaching mathematics, see the following books, available from Dale Seymour Publications:

- *Implementing the* Investigations in Number, Data, and Space® *Curriculum*
- *Beyond Arithmetic: Changing Mathematics in the Elementary Classroom* by Jan Mokros, Susan Jo Russell, and Karen Economopoulos

This book is one of the curriculum units for *Investigations in Number, Data, and Space.* In addition to providing part of a complete mathematics curriculum for your students, this unit offers information to support your own professional development. You, the teacher, are the person who will make this curriculum come alive in the classroom; the book for each unit is your main support system.

Although the curriculum does not include student textbooks, reproducible sheets for student work are provided in the unit and are also available as Student Activity Booklets. Students work actively with objects and experiences in their own environment and with a variety of manipulative materials and technology, rather than with a book of instruction and problems. We strongly recommend use of the overhead projector as a way to present problems, to focus group discussion, and to help students share ideas and strategies.

Ultimately, every teacher will use these investigations in ways that make sense for his or her particular style, the particular group of students, and the constraints and supports of a particular school environment. Each unit offers information and guidance for a wide variety of situations, drawn from our collaborations with many teachers and students over many years. Our goal in this book is to help you, a professional educator, implement this curriculum in a way that will give all your students access to mathematical power.

Investigation Format

The opening two pages of each investigation help you get ready for the work that follows.

What Happens This gives a synopsis of each session or block of sessions.

Mathematical Emphasis This lists the most important ideas and processes students will encounter in this investigation.

What to Plan Ahead of Time These lists alert you to materials to gather, sheets to duplicate, transparencies to make, and anything else you need to do before starting.

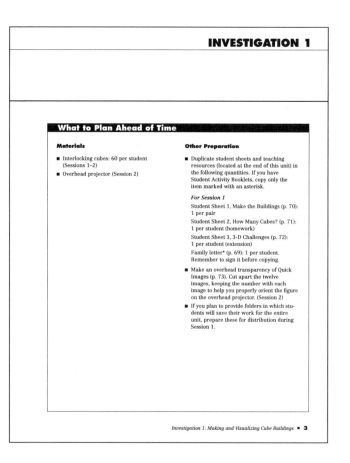

Sessions Within an investigation, the activities are organized by class session, a session being at least a one-hour math class. Sessions are numbered consecutively through an investigation. Often several sessions are grouped together, presenting a block of activities with a single major focus.

When you find a block of sessions presented together—for example, Sessions 1, 2, and 3—read through the entire block first to understand the overall flow and sequence of the activities. Make some preliminary decisions about how you will divide the activities into three sessions for your class, based on what you know about your students. You may need to modify your initial plans as you progress through the activities, and you may want to make notes in the margins of the pages as reminders for the next time you use the unit.

Be sure to read the Session Follow-Up section at the end of the session block to see what homework assignments and extensions are suggested as you make your initial plans.

While you may be used to a curriculum that tells you exactly what each class session should cover, we have found that the teacher is in a better position to make these decisions. Each unit is flexible and may be handled somewhat differently by every teacher. Although we provide guidance for how many sessions a particular group of activities is likely to need, we want you to be active in determining an appropriate pace and the best transition points for your class. It is not unusual for a teacher to spend more or less time than is proposed for the activities.

Ten-Minute Math At the beginning of some sessions, you will find Ten-Minute Math activities. These are designed to be used in tandem with the investigations, but not during the math hour. Rather, we hope you will do them whenever you have a spare 10 minutes—maybe before lunch or recess, or at the end of the day.

Ten-Minute Math offers practice in key concepts, but not always those being covered in the unit. For example, in a unit on using data, Ten-Minute Math must revisit geometric activities done earlier in the year. Complete directions for the suggested activities are included at the end of each unit.

Materials

- Interlocking cubes (60 per student)
- Student Sheet 1 (1 per pair)
- Student Sheet 2 (1 per student, homework)
- Student Sheet 3 (1 per student, extension)
- Family letter (1 per student)

Session 1

Building with Cubes

What Happens

Students put interlocking cubes together to form cube buildings shown in drawings. They verbalize their strategies for building, and compare the sizes of the different structures. Their work focuses on:

- understanding standard drawings of 3-D cube configurations
- exploring spatial relationships between the components of cube configurations

---

Activity

Making Cube Buildings

Give each pair of students 65 interlocking cubes and Student Sheet 1, Make the Buildings. Focus attention on Building 1. Explain that both students in a pair will make their own version of this building with cubes, then compare the two versions.

When they compare their buildings, students are to hold them so that they look just like the drawing. That is, they should be looking straight at the front, right edge, not at the front (the shaded part), the side, or the top.

Building 1

Illustrate (or have students illustrate) this viewing position by standing at the front of the room with your back to the class and with your building positioned so you are looking at the correct edge.

Ask students to describe their strategies for building and to explain how they determined if their buildings were correct. See the **Dialogue Box**, Talking About Cube Buildings (p. 9), for some ideas of what to expect.

Making Buildings 2 Through 8 Students complete the buildings shown in drawings 2–8. Ask them to use a different set of cubes for each building and to keep all their buildings together—this will give you an opportunity to inspect the results. As they work, encourage students to compare their buildings both with the drawings and with buildings made by their classmates.

4 ▪ *Investigation 1: Making and Visualizing Cube Buildings*

Activities The activities include pair and small-group work, individual tasks, and whole-class discussions. In any case, students are seated together, talking and sharing ideas during all work times. Students most often work cooperatively, although each student may record work individually.

Choice Time In most units, some sessions are structured with activity choices. In these cases, students may work simultaneously on different activities focused on the same mathematical ideas. Students choose which activities they want to do, and they cycle through them.

You will need to decide how to set up and introduce these activities and how to let students make their choices. Some teachers present them as station activities, in different parts of the room. Some list the choices on the board as reminders or have students keep their own lists.

Tips for the Linguistically Diverse Classroom At strategic points in each unit, you will find concrete suggestions for simple modifications of the teaching strategies to encourage the participation of all students. Many of these tips offer alternative ways to elicit critical thinking from students at varying levels of English proficiency, as well as from other students who find it difficult to verbalize their thinking.

The tips are supported by suggestions for specific vocabulary work to help ensure that all students can participate fully in the investigations. The Preview for the Linguistically Diverse Classroom lists important words that are assumed as part of the working vocabulary of the unit. Second-language learners will need to become familiar with these words in order to understand the problems and activities they will be doing. These terms can be incorporated into students' second-language work before or during the unit. Activities that can be used to present the words are found in the appendix, Vocabulary Support for Second-Language Learners. In addition, ideas for making connections to students' languages and cultures, included on the Preview page, help the class explore the unit's concepts from a multicultural perspective.

Session Follow-Up: Homework In *Investigations,* homework is an extension of classroom work. Sometimes it offers review and practice of work done in class, sometimes preparation for upcoming activities, and sometimes numerical practice that revisits work in earlier units. Homework plays a role both in supporting students' learning and in helping inform families about the ways in which students in this curriculum work with mathematical ideas.

Depending on your school's homework policies and your own judgment, you may want to assign more homework than is suggested in the units. For this purpose you might use the practice pages, included as blackline masters at the end of this unit, to give students additional work with numbers.

For some homework assignments, you will want to adapt the activity to meet the needs of a variety of students in your class: those with special needs, those ready for more challenge, and second-language learners. You might change the numbers in a problem, make the activity more or less complex, or go through a sample activity with

those who need extra help. You can modify any student sheet for either homework or class use. In particular, making numbers in a problem smaller or larger can make the same basic activity appropriate for a wider range of students.

Another issue to consider is how to handle the homework that students bring back to class—how to recognize the work they have done at home without spending too much time on it. Some teachers hold a short group discussion of different approaches to the assignment; others ask students to share and discuss their work with a neighbor; still others post the homework around the room and give students time to tour it briefly. If you want to keep track of homework students bring in, be sure it ends up in a designated place.

Session Follow-Up: Extensions Sometimes in Session Follow-Up, you will find suggested extension activities. These are opportunities for some or all students to explore a topic in greater depth or in a different context. They are not designed for "fast" students; mathematics is a multifaceted discipline, and different students will want to go further in different investigations. Look for and encourage the sparks of interest and enthusiasm you see in your students, and use the extensions to help them pursue these interests.

Excursions Some of the *Investigations* units include excursions—blocks of activities that could be omitted without harming the integrity of the unit. This is one way of dealing with the great depth and variety of elementary mathematics—much more than a class has time to explore in any one year. Excursions give you the flexibility to make different choices from year to year, doing the excursion in one unit this time, and next year trying another excursion.

Materials

A complete list of the materials needed for teaching this unit follows the unit overview. Some of these materials are available in kits for the *Investigations* curriculum. Individual items can also be purchased from school supply dealers.

Classroom Materials In an active mathematics classroom, certain basic materials should be available at all times: interlocking cubes, pencils, unlined paper, graph paper, calculators, things to count with, and measuring tools. Some activities in this curriculum require scissors and glue sticks or tape. Stick-on notes and large paper are also useful materials throughout.

So that students can independently get what they need at any time, they should know where these materials are kept, how they are stored, and how they are to be returned to the storage area. For example, interlocking cubes are best stored in towers of ten; then, whatever the activity, they should be returned to storage in groups of ten at the end of the hour. You'll find that establishing such routines at the beginning of the year is well worth the time and effort.

Student Sheets and Teaching Resources Student recording sheets and other teaching tools needed for both class and homework are provided as reproducible blackline masters at the end of each unit. We think it's important that students find their own ways of organizing and recording their work. They need to learn how to explain their thinking with both drawings and written words, and how to organize their results so someone else can understand them. For this reason, we deliberately do not provide student sheets for every activity. Regardless of the form in which students do their work, we recommend that they keep their

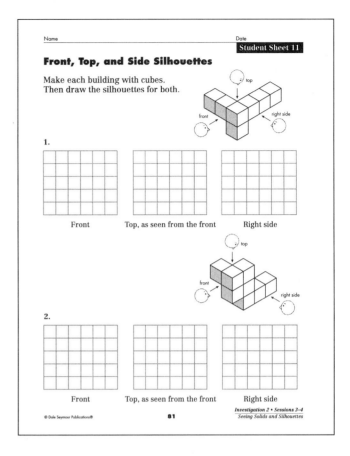

work in a mathematics folder, notebook, or journal so that it is always available to them for reference.

Student Activity Booklets These booklets contain all the sheets each student will need for individual work, freeing you from extensive copying (although you may need or want to copy the occasional teaching resource on transparency film or card stock, or make extra copies of a student sheet).

Calculators and Computers Calculators are used throughout Investigations. Many of the unity recommend that you have at least one calculator for each pair. You will find calculator activities, plus Teacher Notes discussing this important mathematical tool, in an early unit at each grade level. It is assumed that calculators will be readily available for student use.

Computer activities are offered at all grade levels. How you use the computer activities depends on the number of computers you have available. Technology in the Curriculum discusses ways to incorporate the use of calculators and computers into classroom activities.

Children's Literature Each unit offers a list of related children's literature that can be used to support the mathematical ideas in the unit. Sometimes an activity is based on a specific children's book, with suggestions for substitutions where practical. While such activities can be adapted and taught without the book, the literature offers a rich introduction and should be used whenever possible.

Investigations at Home It is a good idea to make your policy on homework explicit to both students and their families when you begin teaching with *Investigations*. How frequently will you be assigning homework? When do you expect homework to be completed and brought back to school? What are your goals in assigning homework? How independent should families expect their children to be? What should the parent's or guardian's role be? The more explicit you can be about your expectations, the better the homework experience will be for everyone.

Investigations at Home (a booklet available separately for each unit, to send home with students) gives you a way to communicate with families about the work students are doing in class. This booklet includes a brief description of every session, a list of the mathematics content emphasized in each investigation, and a discussion of each homework assignment to help families more effectively support their children. Whether or not you are using the *Investigations* at Home booklets, we expect you to make your own choices about homework assignments. Feel free to omit any and to add extra ones you think are appropriate.

Family Letter A letter that you can send home to students' families is included with the blackline masters for each unit. Families need to be informed about the mathematics work in your classroom; they should be encouraged to participate in and support their children's work. A reminder to send home the letter for each unit appears in one of the early investigations. These letters are also available separately in Spanish, Vietnamese, Cantonese, Hmong, and Cambodian.

Help for You, the Teacher

Because we believe strongly that a new curriculum must help teachers think in new ways about mathematics and about their students' mathematical thinking processes, we have included a great deal of material to help you learn more about both.

About the Mathematics in This Unit This introductory section summarizes the critical information about the mathematics you will be teaching. It describes the unit's central mathematical ideas and the ways students will encounter them through the unit's activities.

About the Assessment in This Unit This introductory section highlights Teacher Checkpoints and assessment activities contained in the unit. It offers questions to stimulate your assessment as you observe the development of students' mathematical thinking and learning.

Teacher Notes These reference notes provide practical information about the mathematics you are teaching and about our experience with how students learn. Many of the notes were written in response to actual questions from teachers or to discuss important things we saw happening in the

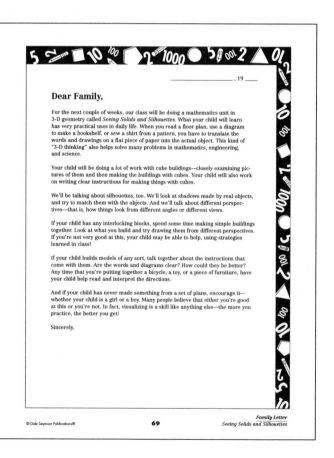

field-test classrooms. Some teachers like to read them all before starting the unit, then review them as they come up in particular investigations.

Dialogue Boxes Sample dialogues demonstrate how students typically express their mathematical ideas, what issues and confusions arise in their thinking, and how some teachers have guided class discussions.

These dialogues are based on the extensive classroom testing of this curriculum; many are word-for-word transcriptions of recorded class discussions. They are not always easy reading; sometimes it may take some effort to unravel what the students are trying to say. But this is the value of these dialogues; they offer good clues to how your students may develop and express their approaches and strategies, helping you prepare for your own class discussions.

Where to Start You may not have time to read everything the first time you use this unit. As a first-time user, you will likely focus on understanding the activities and working them out with your students. Read completely through all the activities before starting to present them. Also read those sections listed in the Contents under the heading Where to Start.

Staying Properly Oriented

◁ Teacher Note

Once students have made a cube building and are trying to draw its silhouettes, they may pick up their building and turn it without keeping track of which side is the front. Then they get confused. After pointing out this potential problem, you might suggest that they keep their building flat on the desktop or table while they move themselves around to view the building from different sides.

You could also have students use a small toy figure to act out viewing the building from different perspectives. Placing a cube building on a table, students should identify the front of the building and draw a "road" passing in front of it (or use a pencil to represent the road). They then place the toy figure looking directly at the front of the building and crouch behind the toy—with their eye level matching that of the toy—to view the building as the toy does.

Next, students make their toy figure "fly" straight up several feet, staying in front of the building, and look down. What does the top of the building look like from this front view? Students then return the toy to its original ground position and make it "walk" around to the right side of the building. They should again crouch to see the right side of the building as the toy figure would.

These activities should help students identify the different views of the cube buildings; a student who is having difficulty may reenact them at any time. However, even when students can correctly identify the different views, they still might not be able to properly visualize the corresponding silhouettes. For such students, always use the overhead to clarify what the building's actual silhouettes look like.

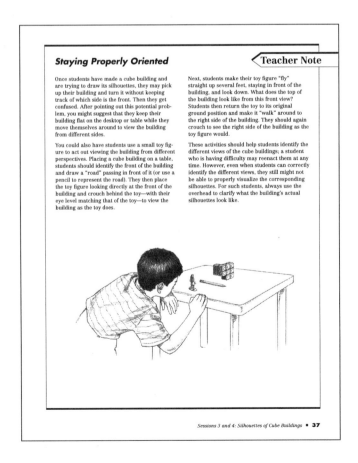

D I A L O G U E B O X

Talking About Cube Buildings

During the activity Making Cube Buildings (p. 4), these students described their strategies for making Building 1.

Building 1

How did you know how to make your building?

Teresa: Make sure it matches up [*to the drawing*]—the number of cubes, the directions.

How do you know if you're right?

Teresa: We counted cubes and compared.

Emilio: It angles here.

B.J.: I looked at the white part on the side first.

Jesse: We each built it and compared.

Show me how to hold your building to compare it to the picture. How did you see the building?

Lina Li: If I hold my building like this, it looks just like the picture.

Karen: My building looks like this. It's not like yours.

Are your buildings different? How are they different?

Lina Li: My building is the same as hers. It's turned. [*Turns Karen's building so its orientation matches that of her building.*]

Karen: OK. They're the same.

The *Investigations* curriculum incorporates the use of two forms of technology in the classroom: calculators and computers. Calculators are assumed to be standard classroom materials, available for student use in any unit. Computers are explicitly linked to one or more units at each grade level; they are used with the unit on 2-D geometry at each grade, as well as with some of the units on measuring, data, and changes.

Using Calculators

In this curriculum, calculators are considered tools for doing mathematics, similar to pattern blocks or interlocking cubes. Just as with other tools, students must learn both *how* to use calculators correctly and *when* they are appropriate to use. This knowledge is crucial for daily life, as calculators are now a standard way of handling numerical operations, both at work and at home.

Using a calculator correctly is not a simple task; it depends on a good knowledge of the four operations and of the number system, so that students can select suitable calculations and also determine what a reasonable result would be. These skills are the basis of any work with numbers, whether or not a calculator is involved.

Unfortunately, calculators are often seen as tools to check computations with, as if other methods are somehow more fallible. Students need to understand that any computational method can be used to check any other; it's just as easy to make a mistake on the calculator as it is to make a mistake on paper or with mental arithmetic. Throughout this curriculum, we encourage students to solve computation problems in more than one way in order to double-check their accuracy. We present mental arithmetic, paper-and-pencil computation, and calculators as three possible approaches.

In this curriculum we also recognize that, despite their importance, calculators are not always appropriate in mathematics instruction. Like any tools, calculators are useful for some tasks but not for others. You will need to make decisions about when to allow students access to calculators and when to ask that they solve problems without them so that they can concentrate on other tools and skills. At times when calculators are or are not appropriate for a particular activity, we make specific recommendations. Help your students develop their own sense of which problems they can tackle with their own reasoning and which ones might be better solved with a combination of their own reasoning and the calculator.

Managing calculators in your classroom so that they are a tool, and not a distraction, requires some planning. When calculators are first introduced, students often want to use them for everything, even problems that can be solved quite simply by other methods. However, once the novelty wears off, students are just as interested in developing their own strategies, especially when these strategies are emphasized and valued in the classroom. Over time, students will come to recognize the ease and value of solving problems mentally, with paper and pencil, or with manipulatives, while also understanding the power of the calculator to facilitate work with larger numbers.

Experience shows that if calculators are available only occasionally, students become excited and distracted when they are permitted to use them. They focus on the tool rather than on the mathematics. In order to learn when calculators are appropriate and when they are not, students must have easy access to them and use them routinely in their work.

If you have a calculator for each student, and if you think your students can accept the responsibility, you might allow them to keep their calculators with the rest of their individual materials, at least for the first few weeks of school. Alternatively, you might store them in boxes on a shelf, number each calculator, and assign a corresponding number to each student. This system can give students a sense of ownership while also helping you keep track of the calculators.

Using Computers

Students can use computers to approach and visualize mathematical situations in new ways. The computer allows students to construct and manipulate geometric shapes, see objects move according to rules they specify, and turn, flip, and repeat a pattern.

This curriculum calls for computers in units where they are a particularly effective tool for learning mathematics content. One unit on 2-D geometry at each of the grades 3–5 includes a core of activities that rely on access to computers, either in the classroom or in a lab. Other units on geometry, measuring, data, and changes include computer activities, but can be taught without them. In these units, however, students' experience is greatly enhanced by computer use.

The following list outlines the recommended use of computers in this curriculum:

Kindergarten
Unit: *Making Shapes and Building Blocks*
   (Exploring Geometry)
Software: *Shapes*
Source: provided with the unit

Grade 1
Unit: *Survey Questions and Secret Rules*
   (Collecting and Sorting Data)
Software: *Tabletop, Jr.*
Source: Broderbund

Unit: *Quilt Squares and Block Towns*
   (2-D and 3-D Geometry)
Software: *Shapes*
Source: provided with the unit

Grade 2
Unit: *Mathematical Thinking at Grade 2*
   (Introduction)
Software: *Shapes*
Source: provided with the unit

Unit: *Shapes, Halves, and Symmetry*
   (Geometry and Fractions)
Software: *Shapes*
Source: provided with the unit

Unit: *How Long? How Far?* (Measuring)
Software: *Geo-Logo*
Source: provided with the unit

Grade 3
Unit: *Flips, Turns, and Area* (2-D Geometry)
Software: *Tumbling Tetrominoes*
Source: provided with the unit

Unit: *Turtle Paths* (2-D Geometry)
Software: *Geo-Logo*
Source: provided with the unit

Grade 4
Unit: *Sunken Ships and Grid Patterns*
   (2-D Geometry)
Software: *Geo-Logo*
Source: provided with the unit

Grade 5
Unit: *Picturing Polygons* (2-D Geometry)
Software: *Geo-Logo*
Source: provided with the unit

Unit: *Patterns of Change* (Tables and Graphs)
Software: *Trips*
Source: provided with the unit

Unit: *Data: Kids, Cats, and Ads* (Statistics)
Software: *Tabletop, Sr.*
Source: Broderbund

The software provided with the *Investigations* units uses the power of the computer to help students explore mathematical ideas and relationships that cannot be explored in the same way with physical materials. With the *Shapes* (grades 1–2) and *Tumbling Tetrominoes* (grade 3) software, students explore symmetry, pattern, rotation and reflection, area, and characteristics of 2-D shapes. With the *Geo-Logo* software (grades 2–5), students investigate rotations and reflections, coordinate geometry, the properties of 2-D shapes, and angles. The *Trips* software (grade 5) is a mathematical exploration of motion in which students run experiments and interpret data presented in graphs and tables.

We suggest that students work in pairs on the computer; this not only maximizes computer resources but also encourages students to consult, monitor, and teach each other. Generally, more than two students at one computer find it difficult to share. Managing access to computers is an issue for every classroom. The curriculum gives you explicit support for setting up a system. The units are structured on the assumption that you have enough computers for half your students to work on the machines in pairs at one time. If you do not have access to that many computers, suggestions are made for structuring class time to use the unit with fewer than five.

Assessment plays a critical role in teaching and learning, and it is an integral part of the *Investigations* curriculum. For a teacher using these units, assessment is an ongoing process. You observe students' discussions and explanations of their strategies on a daily basis and examine their work as it evolves. While students are busy recording and representing their work, working on projects, sharing with partners, and playing mathematical games, you have many opportunities to observe their mathematical thinking. What you learn through observation guides your decisions about how to proceed. In any of the units, you will repeatedly consider questions like these:

- Do students come up with their own strategies for solving problems, or do they expect others to tell them what to do? What do their strategies reveal about their mathematical understanding?

- Do students understand that there are different strategies for solving problems? Do they articulate their strategies and try to understand other students' strategies?

- How effectively do students use materials as tools to help with their mathematical work?

- Do students have effective ideas for keeping track of and recording their work? Do keeping track of and recording their work seem difficult for them?

You will need to develop a comfortable and efficient system for recording and keeping track of your observations. Some teachers keep a clipboard handy and jot notes on a class list or on adhesive labels that are later transferred to student files. Others keep loose-leaf notebooks with a page for each student and make weekly notes about what they have observed in class.

Assessment Tools in the Unit

With the activities in each unit, you will find questions to guide your thinking while observing the students at work. You will also find two built-in assessment tools: Teacher Checkpoints and embedded Assessment activities.

Teacher Checkpoints The designated Teacher Checkpoints in each unit offer a time to "check in" with individual students, watch them at work, and ask questions that illuminate how they are thinking.

At first it may be hard to know what to look for, hard to know what kinds of questions to ask. Students may be reluctant to talk; they may not be accustomed to having the teacher ask them about their work, or they may not know how to explain their thinking. Two important ingredients of this process are asking students open-ended questions about their work and showing genuine interest in how they are approaching the task. When students see that you are interested in their thinking and are counting on them to come up with their own ways of solving problems, they may surprise you with the depth of their understanding.

Teacher Checkpoints also give you the chance to pause in the teaching sequence and reflect on how your class is doing overall. Think about whether you need to adjust your pacing: Are most students fluent with strategies for solving a particular kind of problem? Are they just starting to formulate good strategies? Or are they still struggling with how to start? Depending on what you see as the students work, you may want to spend more time on similar problems, change some of the problems to use smaller numbers, move quickly to more challenging material, modify subsequent activities for some students, work on particular ideas with a small group, or pair students who have good strategies with those who are having more difficulty.

Embedded Assessment Activities Assessment activities embedded in each unit will help you examine specific pieces of student work, figure out what they mean, and provide feedback. From the students' point of view, these assessment activities are no different from any others. Each is a learning experience in and of itself, as well as an opportunity for you to gather evidence about students' mathematical understanding.

The embedded assessment activities sometimes involve writing and reflecting; at other times, a discussion or brief interaction between student and teacher; and in still other instances, the creation and explanation of a product. In most cases, the assessments require that students *show* what they did, *write* or *talk* about it, or do both. Having to explain how they worked through a problem helps students be more focused and clear in their mathematical thinking. It also helps them realize that doing mathematics is a process that may involve tentative starts, revising one's approach, taking different paths, and working through ideas.

Teachers often find the hardest part of assessment to be interpreting their students' work. We provide guidelines to help with that interpretation. If you have used a process approach to teaching writing, the assessment in *Investigations* will seem familiar. For many of the assessment activities, a Teacher Note provides examples of student work and a commentary on what it indicates about student thinking.

Documentation of Student Growth

To form an overall picture of mathematical progress, it is important to document each student's work. Many teachers have students keep their work in folders, notebooks, or journals, and some like to have students summarize their learning in journals at the end of each unit. It's important to document students' progress, and we recommend that you keep a portfolio of selected work for each student, unit by unit, for the entire year. The final activity in each *Investigations* unit, called Choosing Student Work to Save, helps you and the students select representative samples for a record of their work.

This kind of regular documentation helps you synthesize information about each student as a mathematical learner. From different pieces of evidence, you can put together the big picture. This synthesis will be invaluable in thinking about where to go next with a particular child, deciding where more work is needed, or explaining to parents (or other teachers) how a child is doing.

If you use portfolios, you need to collect a good balance of work, yet avoid being swamped with an overwhelming amount of paper. Following are some tips for effective portfolios:

■ Collect a representative sample of work, including some pieces that students themselves select for inclusion in the portfolio. There should be just a few pieces for each unit, showing different kinds of work—some assignments that involve writing as well as some that do not.

■ If students do not date their work, do so yourself so that you can reconstruct the order in which pieces were done.

■ Include your reflections on the work. When you are looking back over the whole year, such comments are reminders of what seemed especially interesting about a particular piece; they can also be helpful to other teachers and to parents. Older students should be encouraged to write their own reflections about their work.

Assessment Overview

There are two places to turn for a preview of the assessment opportunities in each *Investigations* unit. The Assessment Resources column in the unit Overview Chart identifies the Teacher Checkpoints and Assessment activities embedded in each investigation, guidelines for observing the students that appear within classroom activities, and any Teacher Notes and Dialogue Boxes that explain what to look for and what types of student responses you might expect to see in your classroom. Additionally, the section About the Assessment in This Unit gives you a detailed list of questions for each investigation, keyed to the mathematical emphases, to help you observe student growth.

Depending on your situation, you may want to provide additional assessment opportunities. Most of the investigations lend themselves to more frequent assessment, simply by having students do more writing and recording while they are working.

Seeing Solids and Silhouettes

Content of This Unit Students develop spatial visualization skills, a part of geometry that is often neglected in elementary mathematics. They explore ways to pictorially represent solid shapes. They build cube configurations from pictures, mental images, and different types of building instructions. They investigate silhouettes projected by geometric solids and explore what objects look like from different perspectives. Throughout, students learn ways to communicate effectively about three-dimensional objects.

Connections with Other Units If you are doing the full-year *Investigations* curriculum in the suggested sequence for grade 4, this is the third of 11 units. It is one of three in the *Investigations* curriculum that develop students' knowledge of and ability to visualize 3-D geometric objects and contributes to their eventual understanding of volume.

If your school is not doing the full-year program, this unit can also be used successfully in grade 5.

Investigations Curriculum ■ Suggested Grade 4 Sequence

Mathematical Thinking at Grade 4 (Introduction)

Arrays and Shares (Multiplication and Division)

▶ *Seeing Solids and Silhouettes* (3-D Geometry)

Landmarks in the Thousands (The Number System)

Different Shapes, Equal Pieces (Fractions)

The Shape of the Data (Statistics)

Money, Miles, and Large Numbers (Addition and Subtraction)

Changes Over Time (Graphs)

Packages and Groups (Multiplication and Division)

Sunken Ships and Grid Patterns (2-D Geometry)

Three out of Four Like Spaghetti (Data and Fractions)

Investigation 1 ■ Making and Visualizing Cube Buildings

Class Sessions	Activities	Pacing
Session 1 (p. 4) BUILDING WITH CUBES	Making Cube Buildings How Big Are the Cube Buildings? Homework: How Many Cubes? Extension: 3-D Challenges	minimum 1 hr
Session 2 (p. 10) MAKING MENTAL PICTURES	Building from Quick Images Teacher Checkpoint: 3-D Images	minimum 1 hr

Mathematical Emphasis

- Developing concepts and language needed to reflect on and communicate about spatial relationships in 3-D environments

- Understanding standard drawings of 3-D cube configurations

- Exploring spatial relationships between components of 3-D figures

- Developing visualization skills

- Starting to think about problems related to volume

Assessment Resources

Interpreting 2-D Diagrams of 3-D Shapes (Teacher Note, p. 8)

Talking About Cube Buildings (Dialogue Box, p. 9)

Teacher Checkpoint: 3-D Images (p. 12)

Seeing Cube Buildings in Our Minds (Dialogue Box, p. 13)

Materials

Interlocking cubes
Overhead projector
Student Sheets 1–3
Teaching resource sheets
Family letter

Investigation 2 ▪ Exploring Geometric Silhouettes

Class Sessions	Activities	Pacing
Sessions 1 and 2 (Excursion)* (p. 16) SILHOUETTES OF GEOMETRIC SOLIDS	What Are Silhouettes? Predicting the Shape of Silhouettes Matching Solids and Silhouettes What Can You See from Here? Landscapes Challenge Teacher Checkpoint: Match the Silhouettes	minimum 2 hr
Sessions 3 and 4 (p. 30) SILHOUETTES OF CUBE BUILDINGS	Drawing a Building's Silhouettes Drawing All Three Views Assessment: Drawing Before Building Building from Silhouettes Homework: Cube Buildings and Cube Silhouettes Homework: Mystery Silhouettes Extension: Left-Side, Back, and Bottom Silhouettes	minimum 2 hr
Session 5 (Excursion)* (p. 40) DIFFERENT VIEWS OF A CITY	Where Was Each Photo Taken?	minimum 1 hr

◐ **Ten-Minute Math** ▪ **Quick Images**

* Excursions can be omitted without harming the integrity or continuity of the unit, but offer good mathematical work if you have time to include them.

Mathematical Emphasis	Assessment Resources	Materials
▪ Understanding how 3-D geometric solids project shadows with 2-D shapes (for example, how a cone can project a triangular shadow) ▪ Understanding geometric perspective ▪ Learning to visualize objects from different perspectives ▪ Integrating different views of an object to form a mental model of the whole object	Teacher Checkpoint: Match the Silhouettes (p. 25) Difficulties in Visualizing Silhouettes (Teacher Note, p. 26) Good Thinking Doesn't Always Result in "Correct" Answers (Teacher Note, p. 28) Assessment: Drawing Before Building (p. 33) Staying Properly Oriented (Teacher Note, p. 37) Integrating Three Views: How Students Try to Do It (Teacher Note, p. 38) Describing Our Building Silhouettes (Dialogue Box, p. 39)	Overhead projector Geometric solids Toy figures Interlocking cubes Student Sheets 4–18 Teaching resource sheets Plain paper

Investigation 3 ■ "How-To" Instructions for Cube Buildings

Class Sessions	Activities	Pacing
Session 1 (p. 44) WRITING "HOW-TO" INSTRUCTIONS	Writing Good Instructions Homework: More Cube Buildings and Cube Silhouettes	minimum 1 hr
Sessions 2 and 3 (p. 46) WHICH INSTRUCTIONS ARE BEST?	Trying Out Different Instructions Homework: Quick Image Geometric Designs	minimum 2 hr

◗ Ten-Minute Math ■ Quick Images

Mathematical Emphasis

- Interpreting different types of instructions for building with cubes

- Evaluating the effectiveness of different forms of "how-to" instructions

- Developing visualization skills

- Integrating information given in separate views or presented verbally to form one coherent mental model of a cube building

Assessment Resources

Students' Thinking About Instructions (Teacher Note, p. 49)

Materials

Interlocking cubes
Student Sheets 19–25
Teaching resource sheets

Investigation 4 ▪ The Cube Toy Project

Class Sessions	Activities	Pacing
Sessions 1, 2, 3, and 4 (p. 52) MAKING PLANS FOR A CUBE TOY	Building the Toy Making the Instruction Booklet Testing the Plans Assessment: Evaluating the Instruction Booklets Choosing Student Work to Save Homework: Quick Image Dot Patterns Homework: Make a Quick Image Extension: Talent on Display Extension: Wordless Directions	minimum 4 hr

◔ **Ten-Minute Math** ▪ **Quick Images**

Mathematical Emphasis

- Interpreting different types of instructions for building with cubes

- Evaluating the effectiveness of different forms of "how-to" instructions

- Developing visualization skills

- Communicating effectively about three-dimensional objects

Assessment Resources

Assessment: Evaluating the Instruction Booklets (p. 55)

Choosing Student Work to Save (p. 56)

Evaluating Students' Instruction Booklets (Teacher Note, p. 57)

Materials

Interlocking cubes

Student Sheets 26–27

Teaching resource sheets

Following are the basic materials needed for the activities in this unit. Many of the items can be purchased from the publisher, either individually or in the Teacher Resource Package and the Student Materials Kit for grade 4. Detailed information is available on the *Investigations* order form. To obtain this form, call toll-free 1-800-872-1100 and ask for a Dale Seymour customer service representative.

Snap™ Cubes (interlocking cubes): 120 per student. (You can get by with 65 cubes per student, but more will help the final project go more smoothly.)

Resealable plastic bags: 1 per student. Put sets of 120 cubes into each bag.

Wooden geometric solids: ideally, one set for every 6 students. (Note: One set is provided in the materials kit for grade 4; if you have only one set, the activities can be set up at a center.) See p. 17 for a diagram of the needed solids.

Toy figures, 3–4 cm tall: 1 per student pair. Students can bring these from home.

Overhead projector: at least one. If possible, having several projectors would be helpful for students doing the excursions in Investigation 2.

The following materials are provided at the end of this unit as blackline masters. A Student Activity Booklet containing all student sheets and teaching resources needed for individual work is available.

Family Letter (p. 69)

Student Sheets 1–27 (p. 70)

Teaching Resources:

Quick Images (transparency master) (p. 73)

Isometric Grid Drawing (transparency master) (p. 98)

Isometric Grid Paper (p. 99)

One-Centimeter Graph Paper (p. 100)

Three-Quarter-Inch Graph Paper (p. 101)

Quick Image Cubes (transparency master) (p. 102)

Quick Image Geometric Designs (transparency master) (p. 103)

Quick Image Dot Patterns (transparency master) (p. 104)

Practice Pages (p. 105)

Related Children's Literature

Jonas, Ann. *Round Trip.* New York: Greenwillow Books, 1983.

Lord, John. *The Giant Jam Sandwich.* Boston: Houghton Mifflin, 1972.

The M. C. Escher Jigsaw Puzzle Book. New York: Harry N. Abrams, 1996.

Sachs, Michael. *The Pop-up Book of M. C. Escher.* Rohnert Park, Calif.: Pomegranate Press, 1992.

As suggested in the National Council of Teachers of Mathematics *Standards*, visualization skills are an important part of learning in mathematics and science. In fact, several mathematicians and engineers who reviewed this unit commented that it develops skills they have found essential both in college and in their jobs. The major goal of this unit, therefore, is to develop some basic concepts and the language needed to reflect on and communicate about spatial relationships in three dimensional (3-D) environments.

Visualizing a 3-D object by looking at a 2-D drawing is a critical spatial skill. This task is difficult because a 2-D drawing gives us only partial information about the object it represents. For example, the diagram below shows only three sides of the object. Because the picture is two-dimensional, it cannot capture all of the information contained in the actual 3-D object it represents.

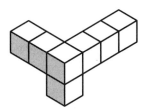

The next diagram, showing three views of the same object, presents even less information.

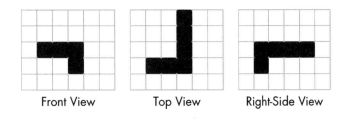

Front View   Top View   Right-Side View

In this type of drawing, not only do students see only three sides, they have no information about how the sides are related or should be put together. To interpret either kind of diagram, students must use the partial information given to form a mental model of the whole object.

Figuring out what the whole object is, given partial information about it, is an important part of problem solving. It requires making inferences, generating and checking possible solutions, and reflecting upon and integrating information gained in previous solution attempts.

Another important notion for both geometry and spatial thinking is that objects look different from different points of view, or perspectives. But students must do more than recognize that this difference exists; they must also learn to visualize how objects look from different views. They develop this skill through problems like the three-views task and problems in which they must figure out what views are possible at different points in given landscapes.

The NCTM *Standards* also recognize the important role of communication in student learning. This unit encourages students to devise ways to communicate effectively about 3-D objects. They explore common graphic methods used to communicate about such objects, for example, isometric and three-views drawings. They also develop suitable vocabulary for describing geometric figures.

To communicate well in writing, students must learn to step into the reader's shoes. In this unit, they see that to write a successful set of building instructions requires that the author see the construction process from the point of view of the reader. Thus, they must learn to read spatial descriptions from the point of view of someone who cannot see the object being described.

Because visualization skills have been neglected by traditional mathematics curricula, many students and adults find such tasks difficult. However, the ability to visualize improves with experience. Through repeated activities with concrete materials, like those provided in this unit, students' visualization skills will gradually increase. Keep in mind, too, that many spatial problems that are difficult to visualize can be easily solved if you build a concrete model of the situation. If, as you are reading the unit, any of the tasks seem difficult, try them yourself using concrete materials, just as the students will be doing.

Because visualization problems involve a set of skills different from those normally taught in mathematics, you may find that some students who normally don't do well in mathematics will excel in this unit. These students have high visualization ability, but not necessarily high language or arithmetic skills. As these students encounter repeated successes, you will see their mathematical self-esteem improve.

At the beginning of each investigation, the Mathematical Emphasis section tells you what is most important for students to learn about during that investigation. Many of these mathematical understandings and processes are difficult and complex. Students gradually learn more and more about each idea over many years of schooling.

Individual students will begin and end the unit with different levels of knowledge and skill, but all will gain greater knowledge about visualizing and describing solids and will develop ways of figuring out the correspondence between 3-D objects and their 2-D representations.

Throughout the *Investigations* curriculum, there are many opportunities for ongoing daily assessment as you observe, listen to, and interact with students at work. In this unit, you will find two Teacher Checkpoints:

Investigation 1, Session 2:
3-D Images (p. 12)

Investigation 2, Sessions 1–2:
Match the Silhouettes (p. 25)

This unit also has two embedded assessment activities:

Investigation 2, Sessions 3–4:
Drawing Before Building (p. 33)

Investigation 4, Sessions 1–4:
Evaluating the Instruction Booklets (p. 55)

In addition, you can use almost any activity in this unit to assess your students' needs and strengths. Listed below are questions to help you focus your observation in each investigation. You may want to keep track of your observations for each student to help you plan your curriculum and monitor students' growth. Suggestions for documenting student growth can be found in the section About Assessment.

Investigation 1: Making and Visualizing Cube Buildings

- What kind of language do students use to communicate about 3-D spatial relationships? How clearly do they describe their construction strategies and make comparisons between buildings?

- How do students understand and interpret drawings of 3-D cube configurations?

- What strategies do students use to keep track of visual information about different parts of 3-D figures?

- When constructing cube buildings from memory, what kinds of mental images do students use to organize their thinking? Do they mentally decompose the building into smaller units, counting cubes in different parts of the building? Do they look for simpler geometric shapes or see the image as a whole object? How many times do students need to see an image flashed before they can successfully reproduce the cube configurations shown?

- How do students predict the number of cubes needed to make a solid rectangular box? Do they think about the box in terms of layers? Do they count only the visible cubes? Do they need to take apart the buildings and count cubes one by one?

Investigation 2: Exploring Geometric Silhouettes

- What expectations do students have about the kinds of 2-D shadows that 3-D geometric solids will project? Do they understand that the shadow projected is not the same as an edge and not always identical to one of the faces of the solid? How do they make sense of discrepancies between their predictions and concrete evidence from holding blocks over the overhead?

- What strategies do students use to match a given view to a given perspective?

- How do students represent cube buildings viewed from different perspectives? Can students accurately draw silhouettes of each side of a cube building from a picture of the building? Do they need to make the building in order to draw the correct silhouettes?

- How do students work with silhouettes of a building viewed from different perspectives to construct the building?

Investigation 3: "How-To" Instructions for Cube Buildings

- How fluently do students interpret different kinds of instructions: 3-D picture, three straight-on views, two straight-on views, layer-by-layer plans, written directions?

- What criteria do students use for evaluating the effectiveness of different instructions? Are they able to take the point of view of another person in assessing the clarity and appropriateness of a set of instructions?

- What strategies do students use to form mental images when moving back and forth between 2-D representations and 3-D constructions?

- How fluently can students put together information about different views of a building to visualize the whole? Are they more comfortable with visual or verbal information?

Investigation 4: The Cube Toy Project

- Which kinds of instructions do students seem most comfortable with? Do they prefer written directions or pictures? Do they prefer pictures that decompose the building into different views or layers? Are they only able to build accurately from a 3-D picture?

- Can students articulate what is good and bad about a set of instructions? Do they recognize that it is important for directions to be both easy to follow and clear enough so that everybody who reads them makes the intended building?

- What kinds of strategies do students use to visualize 3-D buildings from 2-D representations? What kinds of 2-D drawings do they use to represent their 3-D constructions?

- Which methods do students use in their instruction booklets for representing 3-D objects with drawings? Do they use them effectively to write a reasonably clear explanation for building a cube toy? Are they able to take on the reader's perspective?

In the *Investigations* curriculum, mathematical vocabulary is introduced naturally during the activities. We don't ask students to learn definitions of new terms; rather, they come to understand such words as *factor* or *area* or *symmetry* by hearing them used frequently in discussion as they investigate new concepts. This approach is compatible with current theories of second-language acquisition, which emphasize the use of new vocabulary in meaningful contexts while students are actively involved with objects, pictures, and physical movement.

Listed below are some key words used in this unit that will not be new to most English speakers at this age level but may be unfamiliar to students with limited English proficiency. You will want to spend additional time working on these words with your students who are learning English. If your students are working with a second-language teacher, you might enlist your colleague's aid in familiarizing students with these words before and during this unit. In the classroom, look for opportunities for students to hear and use these words. Activities you can use to present the words are given in the appendix, Vocabulary Support for Second-Language Learners (p. 65).

cube building This term refers to any of the various configurations of interlocking cubes that students build and learn to represent in two-dimensional diagrams.

front, top, side, back These terms are used in describing the different *views* of the cube buildings students have made. What does the *front view* of this cube building look like? If this is the *top view* of a cube building, what might it look like from the *side*?

layer (top, middle, bottom) One of the ways of representing a 3-D cube building in a 2-D drawing is to draw each *layer*. Students consider the layer-by-layer approach as one way of making instructions for a cube building.

instructions, directions, Students read and use different types of "how-to" *instructions* or *directions* for building with cubes, then write their own instructions for cube buildings and for a cube toy that they design.

Multicultural Extensions for All Students

Whenever possible, encourage students to share words, objects, customs, or any aspects of daily life from their own cultures and backgrounds that are relevant to the activities in this unit. Since this unit deals largely with geometric objects, there are not many cultural connections. However, when students are designing cube toys in Investigation 4, you might ask them to bring in or describe toys from their own countries or cultures. Perhaps some students can share instruction booklets for building toys or models that are written in another language; this would be a good chance to discuss the benefits of visual instructions.

Investigations

Making and Visualizing Cube Buildings

What Happens

Session 1: Building with Cubes Students put interlocking cubes together to form cube buildings shown in drawings. They verbalize their strategies for building, and compare the sizes of the different structures.

Session 2: Making Mental Pictures After students are briefly shown a picture of a cube building, they construct it from memory by forming and inspecting a mental image of it. This activity gives students experience with visual organization and analysis of images, and more practice communicating about 3-D drawings and structures.

Mathematical Emphasis

- Developing concepts and language needed to reflect on and communicate about spatial relationships in 3-D environments
- Understanding standard drawings of 3-D cube configurations
- Exploring spatial relationships between components of 3-D figures
- Developing visualization skills
- Starting to think about problems related to volume

What to Plan Ahead of Time

Materials

- Interlocking cubes: 60 per student (Sessions 1–2)
- Overhead projector (Session 2)

Other Preparation

- Duplicate student sheets and teaching resources (located at the end of this unit) in the following quantities. If you have Student Activity Booklets, copy only the item marked with an asterisk.

For Session 1

Student Sheet 1, Make the Buildings (p. 70): 1 per pair

Student Sheet 2, How Many Cubes? (p. 71): 1 per student (homework)

Student Sheet 3, 3-D Challenges (p. 72): 1 per student (extension)

Family letter* (p. 69): 1 per student. Remember to sign it before copying.

- Make an overhead transparency of Quick Images (p. 73). Cut apart the twelve images, keeping the number with each image to help you properly orient the figure on the overhead projector. (Session 2)
- If you plan to provide folders in which students will save their work for the entire unit, prepare these for distribution during Session 1.

Building with Cubes

Materials

- Interlocking cubes (60 per student)
- Student Sheet 1 (1 per pair)
- Student Sheet 2 (1 per student, homework)
- Student Sheet 3 (1 per student, extension)
- Family letter (1 per student)

What Happens

Students put interlocking cubes together to form cube buildings shown in drawings. They verbalize their strategies for building, and compare the sizes of the different structures. Their work focuses on:

- understanding standard drawings of 3-D cube configurations
- exploring spatial relationships between the components of cube configurations

Activity

Making Cube Buildings

Give each pair of students 65 interlocking cubes and Student Sheet 1, Make the Buildings. Focus attention on Building 1. Explain that both students in a pair will make their own version of this building with cubes, then compare the two versions.

When they compare their buildings, students are to hold them so that they look just like the drawing. That is, they should be looking straight at the front, right edge, not at the front (the shaded part), the side, or the top.

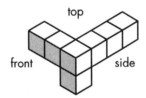

Building 1

Illustrate (or have students illustrate) this viewing position by standing at the front of the room with your back to the class and with your building positioned so you are looking at the correct edge.

Ask students to describe their strategies for building and to explain how they determined if their buildings were correct. See the **Dialogue Box**, Talking About Cube Buildings (p. 9), for some ideas of what to expect.

Making Buildings 2 Through 8 Students complete the buildings shown in drawings 2–8. Ask them to use a different set of cubes for each building and to keep all their buildings together—this will give you an opportunity to inspect the results. As they work, encourage students to compare their buildings both with the drawings and with buildings made by their classmates.

Circulate around the room, stopping here and there to point to a cube in a building and asking the student to show you that cube in the drawing. Such questions will help you assess students' interpretations of the cube drawings and their ability to put together the buildings shown. Can they establish a correspondence between parts of the pictures and parts of the buildings?

When constructing Building 3, many students have trouble correctly placing the cube for the front right leg and make, instead, Building 4 or 5. These students are having difficulty interpreting the drawings; however, their confusion usually clears up as they go on to build all three. See the **Teacher Note**, Interpreting 2-D Diagrams of 3-D shapes (p. 8), for a discussion of some of the difficulties students may have.

Discussing Building Strategies   After students have completed Buildings 1–8, discuss as a class their construction strategies. Ask students to compare the buildings shown in different drawings.

How are Buildings 1 and 2 related? [*They are different pictures of the same cube arrangement*.] **How are Buildings 3, 4, and 5 the same or different?**

❖ **Tip for the Linguistically Diverse Classroom**   To compare the buildings nonverbally, students can point to parts of the buildings that are the same.

How many cubes did you use to make Building 6?

Note that for Building 6, many students extend the back left leg. We can't really tell from the drawing whether there are cubes in the back that are obscured by front cubes. Here, as in other instances in the unit, there is more than one correct building.

As students discuss their interpretations of the drawings, they start to develop the language and concepts needed to communicate spatial information.

To encourage students to start thinking about the size of 3-D objects, ask them to compare the sizes of Buildings 1–8.

Which building is biggest; that is, which uses the most cubes? Which building is smallest, or uses the least cubes? Which is bigger, Building 6 or Building 8? Building 2 or Building 7?

How Big Are the Cube Buildings?

Ask students for their predictions of how many cubes it would take to make the solid, rectangular Buildings 9 and 10. Ask how they made their predictions. Some students use a layer approach. That is, they determine how many cubes are in a vertical or horizontal layer, how many layers there are, then add or multiply to find the total. Other students count only those cubes that are visible in the three exposed sides, and may be counting visible faces rather than cubes.

After discussing their predictions, have students make Buildings 9 and 10 with cubes to check their answers. (To have enough cubes, they will have to take apart Buildings 1–8.)

In determining how many cubes they need for the buildings, some students will have difficulty because they do not organize their counting. These students may double count or miss cubes. Encourage them to use counting strategies they have used in other contexts, such as counting by 2's, 3's, or 4's, or to organize their work by counting the cubes in successive layers.

Also, some students will make the same errors when counting cubes in the actual 3-D buildings as they did when counting them in the drawings. In fact, for some students, determining the number of cubes in rectangular prisms like Buildings 9 and 10 is a very difficult and confusing problem. The only way these students will be convinced of the correct total is to take the buildings apart and count the cubes one by one.

Be sure to follow up with a discussion of the different strategies students used to determine numbers of cubes in the buildings. Let students use any strategy that enables them to count accurately.

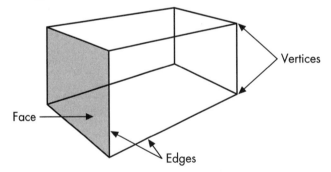

As students talk about the buildings and solids in this unit, they will use their own words (rather than standard terminology) to describe the objects and their parts. For example, they often use the word *corners* instead of *vertices*, the word *sides* instead of *faces*, and the word *boxes* instead of *rectangular prisms*. You can encourage and use such everyday terms yourself, but also introduce and use the standard terms as they fit into the discussion. It is not necessary for students to use the standard terms consistently, but they do need to communicate clearly.

 **Homework**

How Many Cubes? Send home the family letter or *Investigations* at Home booklet with copies of Student Sheet 2, How Many Cubes? Ask students to predict how many cubes are needed to make the buildings shown. Otherwise, they could check their predictions with cubes in class the next day.

Extension

3-D Challenges If your students are ready for an additional challenge, hand out Student Sheet 3, 3-D Challenges. Students can return to these challenges anytime during the unit when they finish an activity early. Challenge 1 will not be difficult for students who have developed a layer strategy for counting cubes.

Challenge 3 asks students if a rather strange looking configuration can be built with cubes. Let the students struggle with this problem. In fact, the figure cannot be built because it is an optical illusion, but let the students come to this conclusion on their own. Some of the students will not understand why this configuration cannot be built; others will be able to address some of the difficulties with the drawing with comments like these:

> It's flat. It's not 3-D.
>
> It does not pop up.
>
> When it goes across, it doesn't go down.
>
> Right down here, it curved and it should be straight.
>
> Can't make it diagonal.
>
> It's impossible.
>
> It's just a picture—that does not mean it really can exist.

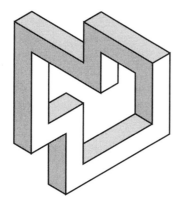

After students have tried this challenge, you might show them some works by the Dutch graphics artist M. C. Escher. Many of his drawings and paintings, like *Waterfall*, depict similar, physically impossible situations.

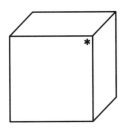

How do you see or interpret this diagram?

Most people see it as a picture of a solid. They think of the vertex marked with the asterisk as protruding toward them; this is the standard interpretation. Other people, however, see the figure as an inside, upper corner of a room. They think of the vertex marked with the asterisk as receding away from them. This is not "wrong," just a nonstandard view.

Some students will have difficulty with the activities at first because they do not see or interpret the diagrams in the standard way. For instance, to make Building 1 on Student Sheet 1, one student started with the building shown at the right below:

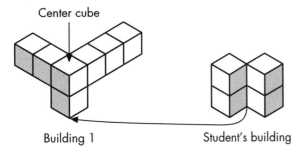

Center cube

Building 1   Student's building

This student was seeing the drawing of the center cube as the inside corner of a room. Thus, she interpreted the entire figure as the upper right-hand corner of a room. As she tried to build what she saw, she constructed the center bottom portion, but then said it was impossible to continue.

Another person, an Investigations staff member, had trouble with Building 8. She saw this figure, which actually contains 8 cubes, as a 5-cube building with a hollow, scoop-like structure in front.

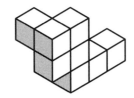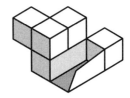

Although most students will "understand" the diagrams of cube configurations, it is not uncommon for some to interpret them in nonstandard ways. Most students with nonstandard interpretations quickly change their ideas when they are made aware that it is not what the artist intended. The student who had trouble with Building 1 changed her interpretation after she heard other students talk about how they saw it. Other students will learn to interpret the diagrams only after spending lots of time building different cube configurations from diagrams. In any case, to help all students better understand these 2-D diagrams, it is essential to have them talk together about their interpretations.

Talking About Cube Buildings

During the activity Making Cube Buildings (p. 4), these students described their strategies for making Building 1.

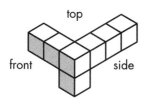

Building 1

How did you know how to make your building?

Teresa: Make sure it matches up [*to the drawing*]—the number of cubes, the directions.

How do you know if you're right?

Teresa: We counted cubes and compared.

Emilio: It angles here.

B.J.: I looked at the white part on the side first.

Jesse: We each built it and compared.

Show me how to hold your building to compare it to the picture. How did you see the building?

Lina Li: If I hold my building like this, it looks just like the picture.

Karen: My building looks like this. It's not like yours.

Are your buildings different? How are they different?

Lina Li: My building is the same as hers. It's turned. [*Turns Karen's building so its orientation matches that of her building.*]

Karen: OK. They're the same.

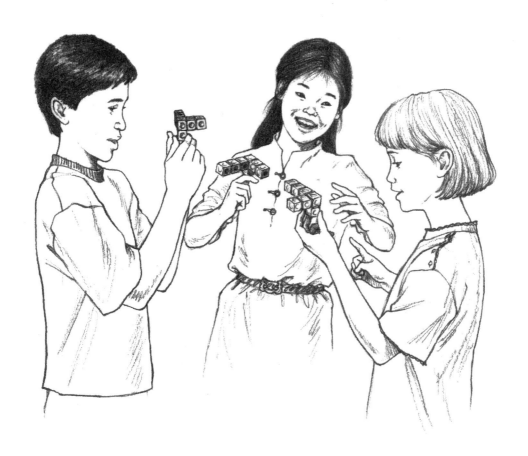

Making Mental Pictures

Materials

- Interlocking cubes (20 per student)
- Overhead projector
- Quick Images transparencies

What Happens

After students are briefly shown a picture of a cube building, they construct it from memory by forming and inspecting a mental image of it. This activity gives students experience with visual organization and analysis of images, and more practice communicating about 3-D drawings and structures. Their work focuses on:

- developing visualization skills
- developing concepts and language needed to reflect on and communicate about spatial relationships

---

Activity

Building from Quick Images

Give each student a supply of 20 interlocking cubes. All students should be seated facing the overhead screen. Explain the activity:

Remember the pictures we used to make cube buildings? Today we're going to try seeing pictures like that in our heads. I will flash a picture of a cube building on the overhead for 3 seconds. Look at it carefully. When I turn off the overhead, try to see the picture in your mind. Then use your cubes to make the building that you saw. I will flash the picture again after everyone has had time to try building it.

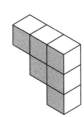

Flash the number 1 picture from Quick Images on the overhead for 3 seconds. It's important to keep the picture up for as close to 3 seconds as possible. If you show the picture too long, students will build from the picture rather than their image of it; if you show it too briefly, they will not have time to form a mental image.

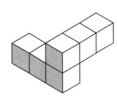 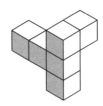 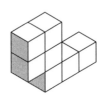

Give the class time to work with their cubes. After you see that most of the building activity has stopped, call students' attention to the overhead and flash the picture again for another 3 seconds. It is essential to provide enough time between the first and second flashes for most students to complete their attempts at building. While they may not have completed their building, they should have done all they can until they see the picture on the screen again.

When the building activity subsides again, show the picture a third time. This time leave it visible, so that all students can complete or revise their solutions.

After students are satisfied that their buildings are complete, ask them to describe how they saw the picture as they looked at it on successive flashes. The **Dialogue Box**, Seeing Cube Buildings in Our Minds (p. 13), demonstrates some typical fourth-grade visualizing strategies.

❖ **Tip for the Linguistically Diverse Classroom** Students can use the Quick Image drawings and hand motions to describe their mental pictures without speaking and writing. The **Dialogue Box** (p. 13) includes a suggestion for melding this approach into the class discussion.

Continue the same activity with additional Quick Images. The transparency provides more pictures than you will need; the more-difficult pictures have higher numbers. Judge from students' progress how difficult the pictures you choose should be.

Quick Images is the theme for the Ten-Minute Math activities suggested for this unit. It is also incorporated as homework in Investigations 3 and 4. Once you have introduced the activity in this session, you can spend 5–10 minutes at various times during the day working on Quick Images, not only for cubes, but also for 2-D geometric designs and dot patterns, as suggested in later investigations.

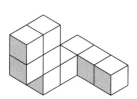
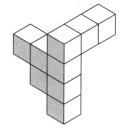
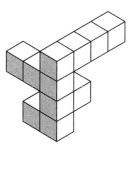
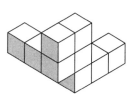

Teacher Checkpoint

3-D Images

Ask students to save the last cube building that they built from a Quick Image picture and to write a description of how they saw it in their mind. Check that the students' buildings are correctly configured and examine their descriptions to see how they are organizing their images. Are they seeing the images as wholes, or as parts, or are they counting parts? See the **Dialogue Box**, Seeing Cube Buildings in Our Minds (p. 13), for some examples of how students think about these images.

Most students should be able to make a Quick Image cube building after two flashes of the picture; all students should be able to build them after the picture is left visible. If you feel your students need additional practice on Quick Image problems, you could make more transparency images from Student Sheet 1, Make the Buildings. Some teachers have found it valuable to repeat some Quick Image pictures that were previously presented.

Seeing Cube Buildings in Our Minds

The class is discussing picture number 2 from the Quick Images transparency.

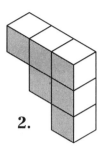

2.

How did you see the cube building in your mind?

Pinsuba: Three cubes going up this way, two this way [*motioning with hands*], one at the bottom.

Nick: I pretended there was an L on the top and a box in the middle.

Kyle: If you turned this upside down, it would be like steps.

Kenyana: If you turn it this way, it looks like a flower.

What did you look for when I showed the picture the second time?

Irena: Where it turns.

Tuong: How many blocks on the arms.

How did you know how many cubes to use for this part [*indicates longer arm*]?

Shiro: I counted.

Rikki: I could just see it.

Note the different organizations students gave to their images. Pinsuba counted cubes in different parts. Nick saw simpler geometric shapes such as an L (a right angle) and a box (a square). Kyle and Kenyana saw the images as whole objects.

It is important for student to discuss their organization strategies. Hearing such strategies described will help students look for similar organization in the next Quick Image picture.

❖ **Tip for the Linguistically Diverse Classroom**
As English-proficient students explain how they saw the cube buildings in their mind, use the transparency to point out (or have the individual students show with their cubes) the way they saw the image.

Do the reverse when limited-English-proficient students offer responses: Have them show with their cubes what they saw in their minds as you articulate their answer. Students could also circle or point to parts of the building on the transparency to show where they focused.

Exploring Geometric Silhouettes

What Happens

Sessions 1 and 2: Silhouettes of Geometric Solids (Excursion)   In the first session, students investigate the shapes of silhouettes projected by geometric solids. In the second session, students build "landscapes" of geometric solids, then identify the locations in each landscape from which pairs of silhouettes could be seen.

Sessions 3 and 4: Silhouettes of Cube Buildings   Students, while looking at pictures of cube buildings, draw silhouettes of the buildings as viewed from the front, top, and right side. They then work in reverse: given front, top, and right-side silhouettes, they construct the matching cube buildings.

Session 5: Different Views of a City (Excursion)   Given a "map" of a cube city with the height of each building marked, students try to identify the locations from which different silhouettes of the city could be seen.

Mathematical Emphasis

- Understanding how 3-D geometric solids project shadows with 2-D shapes (e.g., how a cone can project a triangular shadow)
- Understanding geometric perspective
- Learning to visualize objects from different perspectives
- Integrating different views of an object to form a mental model of the whole object

What to Plan Ahead of Time

Materials

- Overhead projector: at least one; work will go more smoothly if you can get more for students to use in checking their ideas (Sessions 1–4)

- Set of 12 geometric solids (pictured on p. 17): 1 set per 6 students, or 1 set at a center (Sessions 1–2)

- Plain paper: 1 sheet per group (Sessions 1–2)

- For every set of solids, a toy figure 3–4 cm tall: students might bring these from home (Sessions 1–2)

- Interlocking cubes: 30 per pair (Sessions 3–4); 26 per student (Session 5)

Other Preparation

- Number the geometric solids 1 to 12 (you might affix small adhesive labels). Number all sets the same way. Use a different color for each set; this helps in collecting the sets during cleanup. (Sessions 1–2)

- Make and cut apart a transparency of Quick Image Cubes (p. 102) for the associated Ten-Minute Math activity. (Sessions 3–4)

- Duplicate student sheets and teaching resources (located at the end of this unit) in the following quantities. If you have Student Activity Booklets, copy only the item marked with an asterisk.

For Sessions 1–2

Student Sheet 4, Silhouettes of Geometric Solids (p. 74): 1 per pair

Student Sheets 5–7, Landscapes 1–3 (pp. 75–77): 1 per pair

Student Sheet 8, Landscapes Challenge (p. 78): 1 per pair

Student Sheet 9, Match the Silhouettes (p. 79): 1 per student

For Sessions 3–4

Student Sheet 10, Drawing Silhouettes: An Introduction (p. 80): 1 per student, plus 1 transparency*

Student Sheet 11, Front, Top, and Side Silhouettes (p. 81): 1 per student

Student Sheet 12, Drawing Silhouettes A and B (p. 82): 1 per student (cut in half)

Student Sheet 13, Drawing Silhouettes C and D (p. 83): 1 per student (cut in half)

Student Sheet 14, Puzzles: Building from Silhouettes (p. 84): 1 per student

Student Sheet 15, Cube Buildings (p. 85): 1 per student (homework)

Student Sheet 16, Cube Silhouettes (p. 86): 1 per student (homework)

Student Sheet 17, Mystery Silhouettes (p. 87): 1 per student (homework)

For Session 5

Student Sheet 18, Different Views of a City (p. 88): 1 per student

Graph paper that matches the size of your cubes (p. 100 or 101): 1 sheet per student (optional)

Silhouettes of Geometric Solids

Materials

- Overhead projector (at least one)
- Geometric solids (1 set per 6 students)
- Plain sheet of paper (1 sheet per group)
- Toy figures (1 per set of solids)
- Student Sheets 4–8 (1 per pair)
- Student Sheet 9 (1 per student)

What Happens

In the first session, students investigate the shapes of silhouettes projected by geometric solids. In the second session, students build "landscapes" of geometric solids, then identify the locations in each landscape from which pairs of silhouettes could be seen. Their work focuses on:

- understanding how three-dimensional geometric solids project shadows with two-dimensional shapes (for example, how a cone can produce both triangular and circular shadows)
- understanding geometric perspective
- learning to visualize what objects look like from different perspectives

Activity

What Are Silhouettes?

Ask if anyone can tell the class something about silhouettes.

What is a silhouette? How is it produced?

For our purposes, a silhouette is a flat, dark shape that is produced when an object blocks light. We see only the outline of the object. It is like a shadow.

Illustrate how the silhouette of your hand is formed when you place it on the glass screen of an overhead projector. The light coming from the projector is blocked by your hand. The mirror above your hand bends the light so that the silhouette projects onto the screen or wall instead of the ceiling.

If there's time outside of the math hour, students enjoy making standard above-the-neck silhouettes. You can do these quite easily with a stable light source (such as a reading lamp) in a darkened room, with students sitting in front of a piece of paper taped to a wall. Trace their shadow, and let them cut it out.

Predicting the Shape of Silhouettes

Introduce the set of 12 geometric solids, asking students to name any they know. They should be familiar with some of the names, but will be unfamiliar with many others and may offer the names of two-dimensional shapes instead.

You should use the correct names and encourage the students to do so, but do not give definitions for the solids nor require students to memorize the names.

Note: The rectangular prism in this set of solids is very close to the cube in its dimensions. Drawings of it may be confused with those of the cubes.

Hold up the large cube from the geometric solids set.

Look at this cube. What do you think its silhouette will look like when I put it on the overhead, like this?

With the overhead projector's light off, place the cube on the glass so that it is directly underneath the mirror, and the face you showed the students is facing down. Have the students draw their predictions. Then turn on the overhead.

Does your prediction look like the actual silhouette? If not, how is it different?

Repeat this procedure with the narrow cylinder. Show students the circular base of the cylinder, then place the base on the overhead glass (with the projector light off). Ask students to draw their predicted silhouettes, and walk around the classroom to see what they draw. After each student has made a prediction, turn on the overhead so that students can discuss the actual shape.

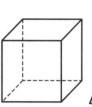 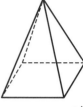
cube   square pyramid

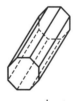
octagonal prism   cylinder

rectangular prism   cone

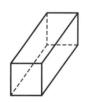
square prism   hexagonal prism

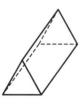
triangular prism   cylinder

sphere   hemisphere

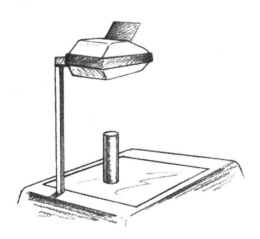

Next, ask them to predict the silhouette shape of the cylinder's curved side. Hold up the cylinder, with its top and bottom horizontal, showing the long, curved side. Then lay the cylinder on its side on the overhead projector glass (light off).

Again, as students draw their predictions, walk around the classroom to see their ideas. After everyone has made a prediction, turn on the overhead. This problem is a bit more difficult, because many people (both students and adults) do not expect a curved figure to produce a rectangular silhouette; they have not had much experience with this type of visualization.

Can you find something else in the room that has a rectangular silhouette? What will the silhouette of a piece of chalk [*placed on its side on the overhead glass*] look like?

You can test students' predictions by putting the objects they suggest on the overhead.

Activity

Matching Solids and Silhouettes

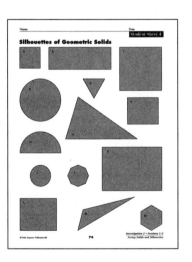

Give one copy of Student Sheet 4, Silhouettes of Geometric Solids, to each pair of students, or place copies of this sheet at a center with your set of wooden solids.

Look at each silhouette on the sheet, and try to find all the solids in our set that could make that shape. Beside or below each silhouette, write the number of every solid you think could make it. If you think that none of our solids would make a particular silhouette, write "none" below it.

❖ **Tip for the Linguistically Diverse Classroom** Supplement your verbal instructions for Student Sheet 4 with nonverbal cues, as follows:

Point to silhouette A and then to any of the solids. Students are to give thumbs up if they believe the solid will produce silhouette A, thumbs down if not. If they indicate that it *will* make the silhouette, write the number of that solid next to silhouette A. Do this for each of the remaining solids.

To demonstrate when a student might write "none," draw a T-shaped silhouette on the board. After pointing to each solid and giving the thumbs down sign, write the word *none* next to the T-shaped silhouette.

Tell students that the silhouettes on the sheet are very nearly the same size as the solids and that size is important in making their matches.

Note: Because of manufacturing variance, the solids and the silhouettes may have slightly different sizes. Compare your solids with the silhouettes to see if there are size discrepancies; if there are, make allowances for "mistakes" that may be due to these discrepancies rather than to conceptual problems.

Checking Student Predictions Most students will use the solids to check their predictions. Some will place the solids directly on the student sheet; others will want to use the overhead to see the actual shadows cast. Circulate to observe students' progress. The **Teacher Note**, Difficulties in Visualizing Silhouettes (p. 26), describes several common difficulties students encounter and tells how to help students with them.

The overhead projector is an essential tool for helping students understand silhouettes. It provides silhouettes that students can actually see rather than merely imagine; it gives them something "concrete" that they can manipulate. You should repeatedly urge students to "prove" their answers by using the overhead.

The following list shows reasonable answers for this activity. However, any answer that students can justify or "prove" should be considered "right."

Silhouette	*Solids that can make the silhouette*
A	square prism
B	square prism, triangular prism, narrow cylinder, octagonal prism, hexagonal prism (may be too large)
C	large cube, wide cylinder, square pyramid
D	sphere, hemisphere, wide cylinder
E	triangular prism
F	rectangular prism
G	hemisphere
H	square pyramid, cone
I	narrow cylinder
J	octagonal prism
K	large cube (viewed looking straight on at one edge)
L	rectangular prism
M	none
N	hexagonal prism

What Can You See from Here?

Many solids can make more than one silhouette. Who can show us a solid that will make several silhouettes with different shapes? What shapes will it make? Are there other solids that will make several silhouette shapes? How could one solid make different shapes of silhouettes?

Students should understand that as we look at a solid from different positions, we might see different silhouettes. For example, looking at a cylinder from the end, we see a circular silhouette; from the side, a rectangular shape.

Divide your class into as many groups as you have sets of geometric solids. Give each group a set of solids and a blank sheet of letter-size paper. Distribute Student Sheet 5, Landscape 1, to every pair of students within the working groups. If you have just one set of solids, set up the activity at a center where small groups can work.

The diagram on Student Sheet 5 shows a particular landscape, or arrangement, of nine geometric solids. Each group is going to build this arrangement. Place your solids on a sheet of plain paper exactly as shown in the diagram. (Be careful when you place the sphere so that it doesn't roll out of position.) When you are done, everyone in the group should agree that your landscape matches the diagram.

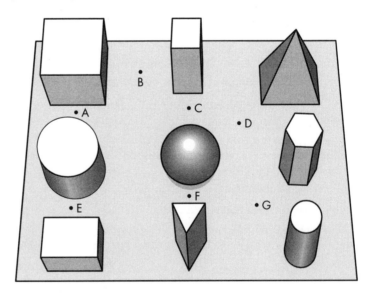

It's important to have students set up their solids on this standard-size paper, because their answers may differ if the distances between the solids vary.

When the landscapes are built, have each group of students place a toy figure at point B, referring to the diagram to find that position.

Let's suppose that your little person is walking around in this geometric landscape. The person is standing at point B. What will the silhouette of the square prism look like to this person? [*Hold up the prism in an upright position.*] **Draw what you think it will look like.**

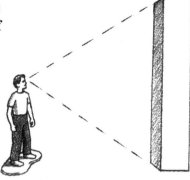

After students make a prediction, show the silhouette on the overhead by placing the prism on the overhead glass as shown in the illustration.

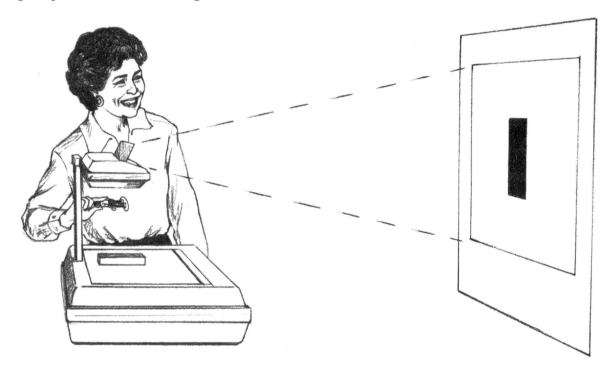

Note: In order to show the proper silhouette, you must change the positioning of the figure from vertical to horizontal. If students question your positioning, explain that the mirror on the projector is like the eyes of the little person looking at the square prism. To help them understand the change, it may help to position the toy figure up underneath the mirror with its feet toward the projected image and looking directly down at the prism.

This change of perspective is difficult for students to understand, so don't belabor the point. You will have to be the official "positioning person" of objects on the overhead.

Does this silhouette look like your prediction? Why or why not?

Next, direct attention back to Student Sheet 5:

Look at the first pair of silhouettes below the diagram for Landscape 1. Find a lettered point in the diagram where your little person could be standing and see both of these silhouettes. Whatever point you want to try, the little person could be facing in any direction.

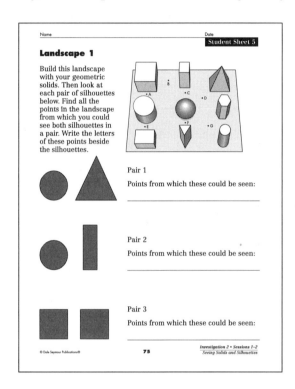

If you think you know an answer, put your toy figure at that point in your landscape and try to convince the others in your group that someone could see both silhouettes from that position. You might have to crouch down and get behind your figure to see the solids the way it does. Remember, it's not only the shape of the silhouettes that is important. The size matters, too. If the silhouette shows a large square, you must find a point from which you would see a large square silhouette.

❖ **Tip for the Linguistically Diverse Classroom** As you give oral directions for Student Sheet 5, simultaneously use appropriate actions. Point to a pair of silhouettes on the sheet. Then place a toy figure somewhere in the landscape. Crouch down behind the figure. Look around, peering at the surrounding shapes. Look confused, then change the placement of the figure. Crouch down, look around again, and nod vigorously. Point to the students, asking for a thumbs-up or thumbs-down signal to check if they are in agreement.

Judging Student Answers   Some students might say that the answer to this first problem is point D; others might say point C. Both answers are reasonable. In fact, there are several reasonable answers for most pairs of silhouettes (see the list at the end of this activity). Some students might select point G; but notice that from point G, the figure will not get an unobstructed view of the entire pyramid. Some students will not understand what's wrong with point G unless they position the toy figure at G—facing toward the square pyramid—and crouch behind the toy to see exactly what it is "seeing."

After pairs of students complete on their own the next two pairs of silhouettes, discuss them as a class. For each problem, students must justify their answers. Encourage other students to challenge any answers they believe are incorrect. The class should discuss disagreements until a consensus is reached.

Students can find, test, and justify their answers by moving the toy figure around to different points in the solids landscape, then getting behind it to "see as it sees." You might help resolve some controversies by positioning the solids in question on the overhead projector.

Landscapes 2 and 3   Continue the activity with Student Sheet 6, Landscape 2. Students work in their small groups to build this new landscape, then in pairs as they complete the student sheet. Discuss their answers as a class. Continue with Student Sheet 7, Landscape 3. This time you might challenge students to try to figure out the answers without first building the landscape with the solids. They can then check their answers by building.

Almost all students will be able to successfully complete the landscape problems by using solids, and many will be able to solve at least some of the problems without the solids. Be sure that students keep Student Sheets 5–7 for use in the next activity.

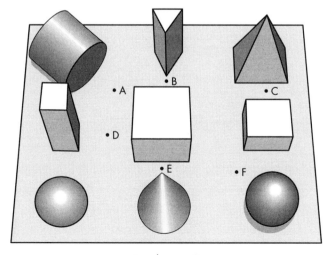

Landscape 2

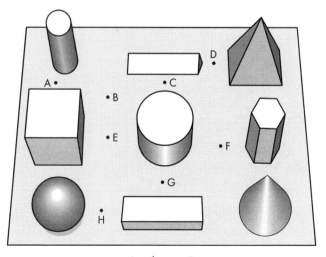

Landscape 3

Reasonable Answers The following chart lists reasonable answers for Student Sheets 5–7. We suggest that you let students decide which answers they will accept as correct; they sometimes can make a good case for answers that are not listed below. Watch for and encourage good reasoning in the student discussions, even if it doesn't lead to "correct" answers. (See the **Teacher Note**, Good Thinking Doesn't Always Result in "Correct" Answers, p. 28.)

Student Sheet 5, Landscape 1	Student Sheet 6, Landscape 2	Student Sheet 7, Landscape 3
Pair 1: C, D Pair 2: B, C, F, G Pair 3: A, C	Pair 1: A, F Pair 2: E Pair 3: C	Pair 1: D Pair 2: H Pair 3: C, G

Activity

Landscapes Challenge

Make available Student Sheet 8, Landscapes Challenge, to those who would enjoy an additional challenge. This problem is like those on Student Sheets 5–7 (which students will need for this activity), except that the given silhouette pair could have been seen in any or all of the landscapes.

Because students can build only one landscape at a time, this activity encourages them to make their predictions by examining the diagrams. Of course, they can check their answers by actually building the landscapes with the geometric solids again.

Following are reasonable answers for the Landscapes Challenge. Be sure to allow students to discuss and defend any alternative answers they have found.

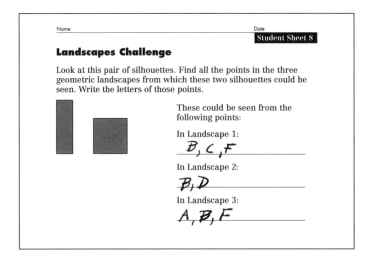

Student Sheet 9, Match the Silhouettes, can be used as part of your ongoing assessment of students' progress. Students are to match the silhouette to a picture of the geometric solid that made it. If needed, students should be allowed to use solids to help them identify silhouettes (but note that the solids in the pictures do not correspond to solids in the set you have been using). If students do use solids, they should say so on their sheet.

Geometric Solids

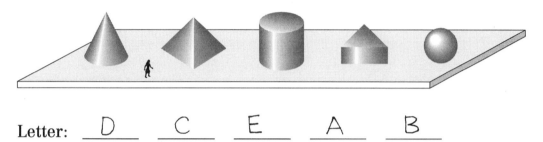

Letter: __D__ __C__ __E__ __A__ __B__

Silhouettes

Have students hand in their completed sheets. Note how many students make errors and what types of errors they make. For instance, some students will match the silhouettes to the top views of the solids, not taking the perspective of the artist who is viewing the solids from the side.

Also, note which students used solids and which students were able to visualize the correct silhouettes without using the solids. Either way, students who get the correct answers are making good progress in this sort of geometry.

Finally, have the students discuss their answers in a whole-class session. Have them justify their answers:

Why did you say that the shadow for the pyramid was triangle C?

Difficulties in Visualizing Silhouettes

Many students do not understand exactly what part of a solid is producing the shape of its silhouette. In some cases—depending on the viewing angle—a silhouette is identical to one of the faces of the solid; in other cases, it is not.

For instance, the octagonal prism can make this rectangular silhouette.

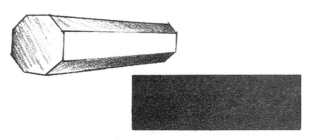

Some students think that the silhouette should be considerably narrower—that it would be identical to one of the prism's rectangular faces. Try this yourself to see the difference.

Other students imagine the silhouette of a cube viewed diagonally as a line segment (one of its edges) or a square (one of its faces), rather than a rectangle. To demonstrate that it is a rectangle, place a cube on the overhead with only one of its edges touching the glass.

One way to help students is to ask them to predict the silhouette for a solid by drawing its outline on an overhead transparency. For example, if students predict that the silhouette for the hexagonal prism is produced by one of its faces, they would trace this face on a transparency. They will see their mistake when they place the transparency, along with the solid, on the overhead and compare their drawing to the actual silhouette. The resulting discrepancy will cause them to reevaluate their theory of silhouettes.

As you are showing silhouettes on the overhead, many students will notice a bit of "overhang."

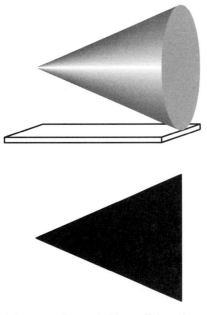

Silhouette of cone held parallel to glass

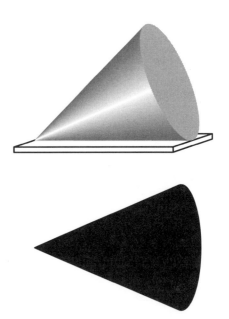

Silhouette of cone laid flat on glass

Continued on next page

For instance, if you place the narrow, square prism with one of its square bases on the glass, you might (depending on where the prism is positioned) see not only the silhouette of the base, but parts of the long rectangular faces. Acknowledge this, but tell students that we are trying to think about "perfect" silhouettes—what we would see if we took these objects outside on a sunny day when the sun is directly overhead.

Some silhouettes can be produced only by positioning solids on the overhead in specific ways. For example, the cone produces a triangle only if held parallel to the overhead glass (see diagram below).

Some students will lay the cone flat on the overhead and claim that it cannot make a triangular silhouette. Most are easily convinced otherwise by a demonstration of the parallel position. For some students, however, the problem lies

deeper. For example, here's how one student thought about the silhouette of a cone as he tried to match it to the shapes on Student Sheet 4:

Alex: The cone doesn't match any of them [the silhouettes on Student Sheet 4] because it's curved on the bottom.

What about now? [*Teacher holds the cone parallel to the overhead.*]

Alex: Nope, because I still know it's curved. It's the same thing.

The difficulty here is not simply that the student is confused by the circular overhang. It is a conceptual problem—the student can't believe that a circular object can produce a straight-sided silhouette such as a triangle. This student needs more experience viewing silhouettes on the overhead, or in direct sunlight.

Good Thinking Doesn't Always Result in "Correct" Answers

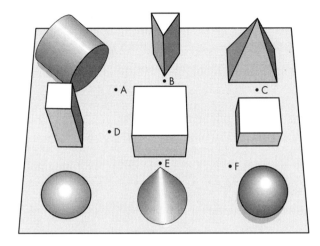

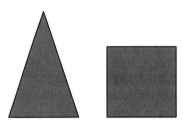

Pair 2

A class has been working on Student Sheet 6, Landscape 2. Now they are talking about silhouette pair 2 (as shown above). They are trying to decide from which points in the landscape this pair of silhouettes can be seen.

David: I think point B works because you can see the front of the square, and if you tilt your head up you can see a triangle—but you don't see the whole entire triangle, just the top of the triangle [*he draws a triangle in the air*].

Can you show us that on Landscape 2?

David places a tiny upright figure at point B in the landscape that his group built from the wooden solids. The figure directly faces the cube in the center.

David: OK, point B is right here. See, if you focus, you can see this [*moves his finger from the eyes of the figure to the closest face of the cube*], you can see the square. Then, if you tilt your head up, you can see the top of the triangle [*he places one finger horizontally across the top of the cone, about three-quarters of the way up from the bottom, forming a triangle*].

The top of the cone?

David: Yeah, that's a triangle.

That's true. But it's not the whole triangle we see in the silhouette pair—is that OK?

David: Yes, we're just supposed to see a triangle.

Class, what do you think about David's reasoning? Do you believe it?

Most of the class members say yes. In a strict mathematical sense, there are several problems with David's arguments. First, if the toy figure viewed the cube and the upper portion of the cone from point B, it would not see the silhouettes in pair 2; instead, it would see a triangle much smaller than the square. Second, it's not clear that David's toy figure has a line of sight that includes the top of the cone; it depends on the size of the toy figure and its position among the solids.

However, we see much more that is correct with David's reasoning than is incorrect. First of all, he definitely understands the idea of perspective. He knows where point B is located in the landscape, and he is able to project himself into the position of the toy figure and imagine what it sees. He has thus achieved one of the major goals of this session. Further, he clearly explains and communicates his thinking to his classmates. His argument is strong enough that he convinces his classmates that he is correct.

Continued on next page

Effective communication about spatial relationships is another major goal of this unit. Thus, we should be very pleased with David's reasoning. Even though he has made some "mistakes," he has clearly taken the initial steps on the long, difficult journey of understanding and visualizing different geometric perspectives.

Recognizing that students' thinking could be even sharper, the teacher tries to get the class to think a bit more carefully about the situation:

OK, everyone agrees that our toy figure will see a triangle if it sees the top of the cone. But how can we be sure that it sees the cone? How can we tell for sure?

David: Look, it sees from here to here [*quickly drawing a line in the air from the toy's eyes to the upper portion of the cone*].

Irena: Wait a minute. [*She goes around the table and crouches directly behind the toy figure, with her eyes at the same level as those of the toy.*] I can't see the cone.

David: Come on, let me see. [*He gets in the same position as Irena.*] I see it, look.

Irena [*returning to her crouching position.*]: Well, I see it if I put my head this high, but I don't see it if I put my head lower.

What do the rest of you think?

After several more students take the viewing position of Irena, the class decides that you might see the triangle or you might not, depending on how high your eyes are. They conclude that David's answer and reasoning should be considered correct. The teacher then returns to the issue of size.

While you were stooping down and looking at the solids, how did the sizes of the square and triangle compare?

Nhat: They were different. The triangle was small.

So does the toy figure see the triangle and square in pair 2?

David: Yes, it is a triangle.

Nhat: But it's a small triangle. It's not like the picture.

Most, but not all, of the students say that the triangle is too small, that you can't really see the large triangle silhouette in pair 2 by looking at the cone from point B. The teacher decides not to belabor the issue at this time, knowing that there will be opportunities to reconsider it.

Many students at this age level do not attend to the size factor in making judgments about these landscape silhouettes. It is usually best to do as this teacher did—that is, raise the issue, but avoid imposing a "correct" answer on the class.

Silhouettes of Cube Buildings

Materials

- Overhead projector (at least one)

- Interlocking cubes (30 per pair)

- Transparency of Student Sheet 10

- Student Sheets 10–11 (1 per student)

- Student Sheets 12–13 (1 per student, cut in half)

- Student Sheet 14 (1 per student)

- Student Sheets 15–17 (1 per student, homework)

What Happens

Students, while looking at pictures of cube buildings, draw silhouettes of the buildings as viewed from the front, top, and right side. They then work in reverse: given front, top, and right-side silhouettes, they construct the matching cube buildings. Their work focuses on:

- understanding geometric perspective

- learning to visualize what objects look like from different perspectives

- integrating different views of an object to form a mental model of the whole object

Note: If you have skipped the two excursion sessions at the beginning of this investigation, read the activity What Are Silhouettes? (p. 16). Introducing and defining silhouettes will probably add about 5–10 minutes to the session.

 Ten-Minute Math: Quick Images   Once or twice during the next few days, do the Quick Images activity. Use the pictures of cube buildings. Remember, Ten-Minute Math is to be done outside of the mathematics hour.

Students will need about 20 cubes. If you place buckets of cubes at several locations around the classroom, students can quickly get what they need. Choose two or three pictures of cube buildings from the Quick Image Cubes transparency.

Flash one picture for 3 seconds; students then try to make the building.

Flash it for another 3 seconds, and let students revise their work; then leave the picture visible for a final revision.

After finishing a building, students talk about what they saw in successive flashes.

For variations and a review of the procedure for Quick Images, see p. 63.

Distribute Student Sheet 10, Drawing Silhouettes: An Introduction. First have students use cubes to make the building shown. Then show the transparency of this sheet on the overhead. Explain that two silhouettes of this cube building are drawn on the grids. One was seen by looking at the front of the building; the other was seen by looking at the right side of the same building. Point out that each square on the grid corresponds to one cube.

Drawing a Building's Silhouettes

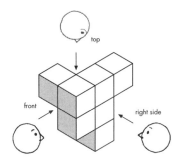

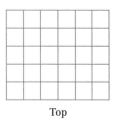

Front

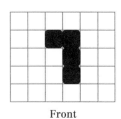

Top

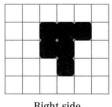

Right side

Ask a volunteer to show the class (from the front of the classroom) the front of this cube building. Then place the student's building on the overhead glass so that the front of the building is facing upward, directly under the mirror (see the diagram). Check the silhouette on the overhead screen, matching it with the grid drawing. Support the building with your hand as necessary. Repeat this procedure for the right-side silhouette.

Silhouettes from the Top Have students position their buildings exactly as shown on Student Sheet 10. Ask them to draw the top silhouette—the shape seen by looking down at the top of the building—in the remaining grid. Look for these four positions of the silhouette in student work, and recreate them on the board or overhead for class discussion:

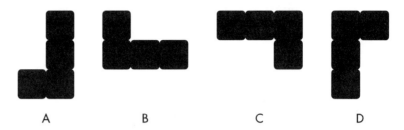

A   B   C   D

Ask students how these four silhouettes are different and how they are the same. You can demonstrate on the overhead, by turning the building, that each is a top silhouette seen from a different side of the building.

Point out that in diagrams like this, to avoid confusion, we usually draw the top silhouette as seen from the *front* side [figure A, above].

Activity

Drawing All Three Views

Distribute Student Sheet 11, Front, Top, and Side Silhouettes. Students first use cubes to make each cube building shown, then draw its front, top, and right-side silhouettes.

Remind students to follow the standard practice of drawing the top as viewed from the front. See the **Teacher Note**, Staying Properly Oriented (p. 37), for ideas on helping students keep track of which side is which. Pairs of students should compare answers and discuss any disagreements.

When everyone has completed Student Sheet 11, have students discuss the solutions. See the **Dialogue Box**, Describing Our Building Silhouettes (p. 39), for a typical student discussion of the front and top silhouette for Building 2 on the student sheet. This discussion demonstrates two difficulties that commonly arise—drawing the top view from other than the front, and failing to draw recessed cubes. As in this example dialogue, some disagreements may be resolved only by positioning cube buildings on the overhead to show their actual silhouettes.

Again, it is essential to use the overhead projector throughout this session; without it, many students will be unable to visualize the different views. The overhead projector provides silhouettes that students can actually see and manipulate. For instance, when looking at a cube building that has several recessed cubes, many students imagine that these cubes cannot be seen. It is far more effective to show the actual silhouette on the overhead and point out where the recessed cubes show up than to present a logical argument about why the cubes should appear in the silhouette.

Assessment

Drawing Before Building

Student Sheets 12 and 13 can help you assess students' strengths and weaknesses in their visualization strategies. In this activity, you learn whether students, given a picture of a cube building, can accurately draw the front, side, and top silhouettes without making the buildings, and if they cannot do it from a picture, whether they can accurately draw these silhouettes by first making the buildings with cubes.

Use Student Sheet 12, Drawing Silhouettes A and B, to introduce the task and to informally observe student progress in class; then use Student Sheet 13, Drawing Silhouettes C and D, to collect information that you can analyze more carefully.

Student Sheets 12 and 13 should be cut in half. Distribute them sequentially, a half sheet at a time, starting with problem A.

For problem A, draw what you think the silhouettes of this building look like. Don't use your cubes as you make these drawings.

After students have completed their drawings, distribute the half sheet with problem B.

This is the same building you just worked with. Make it with cubes, then draw the silhouettes again.

When they have finished drawing, ask:

Are these silhouettes the same as the ones you made for problem A? How are they different? Was it easier to make the silhouettes with or without the actual cube building? Why?

Repeat this procedure with the half sheets of Student Sheet 13, Drawing Silhouettes C and D.

What to Look For   For problems A and B, it is sufficient to walk around the classroom and observe students' work. For problems C and D, collect students' work so you may examine it more carefully later. We have deliberately not provided answers for these sheets; it's important that you try the problems yourself to understand the processes that students must go through to solve them.

Once students actually construct the buildings with cubes (problems B and D), the answers should be clear. As you examine the solutions for both parts of each problem, note whether students have drawn the correct silhouettes; if not, try to notice what is difficult for them. Common errors include interchanging views (that is, drawing the side for the front), not drawing cubes that are recessed, and drawing a top view that is not from the front perspective (for example, from the side). Your assessment should note (1) whether students were able to draw the silhouettes accurately, and (2) whether they needed to make the building in order to draw the correct silhouettes.

- Students who can draw correct silhouettes, whether they need to use the actual cube buildings or not, are developing both good visualization skills and an understanding of how two-dimensional pictures represent three-dimensional objects.

- Those who cannot draw the silhouettes without making the buildings first need more experience trying to draw the three views from images, then checking their answers by building. The repeated comparisons of their images to the buildings will help them sharpen their imagery skills.

- Students who cannot draw the correct silhouettes even after making the buildings need more experience with making cube buildings, then drawing the three views.

Activity

Building from Silhouettes

Hand out Student Sheet 14, Puzzles: Building from Silhouettes. Explain that the students are to construct cube buildings that make the three given silhouettes.

Though these problems are difficult, they foster the development of an extremely important mental skill, the integration of several separate views into a single concept of the whole object. For more information on this skill and how you will see it developing in your students, read the **Teacher Note**, Integrating Three Views: How Students Try to Do It (p. 38).

Because of the difficulty of these problems, you will see a variety of unconventional solution attempts. For instance, some students will make a different building for each of the three silhouettes. These students need to be reminded that they are looking for one building that will make all three silhouettes.

Other students might try to solve the problem by rearranging the cubes in their building for each different view. That is, they will show you the front view of their building; but then, when they show you the top view, they will move some of the cubes. Remind them that they are looking for just one building that works for all views—rearranging the cubes changes the building into a different one.

Reassure students that these problems, like other sorts of puzzles, may take some time to solve. The answers are not supposed to be found easily or quickly. Such comments can help students with low mathematical self-esteem develop proper expectations about their performance. If they think that the answers should come quickly and easily, they may become frustrated. But if they view the problems as challenging puzzles to solve, they will be more inclined to persevere in their search for answers, and will be justifiably proud when they find solutions.

The challenge at the bottom of the puzzle sheet is to find as many buildings as possible that will make each set of silhouettes. For the first two puzzles, there are several possible buildings. For puzzle 1, for example, one simple solution is a $2 \times 2 \times 2$ cube; another solution would be the same cube with any one cube removed. You could get a third solution by removing (from a $2 \times 2 \times 2$ cube) two cubes that do not have adjacent faces (there are several ways to do this, as you will see if you try it).

1.   Front   Top, as seen from the front   Right side

You should know that for puzzle 3 only one building will work. When students conclude that there is only one building for the third puzzle, ask them to "prove" it.

🏠 Homework

Cube Buildings and Cube Silhouettes   After Session 3, send home copies of Student Sheet 15, Cube Buildings, and Student Sheet 16, Cube Silhouettes. Students draw the front, top, and side-view silhouettes of each of the buildings. Students could check their answers with cubes in class the next day.

Mystery Silhouettes   Students find objects at home that will make silhouettes with given shapes. They list the objects they find on Student Sheet 17, Mystery Silhouettes. On the back of the sheet or on a separate piece of paper, students draw a silhouette of some object they find at home; in class, other students try to guess what the object is.

❖ **Tip for the Linguistically Diverse Classroom**   To clarify this homework assignment, draw a silhouette of a simple classroom object, such as the globe. Write a question mark beside the silhouette. Pointing to various objects in the room, shrug your shoulders to indicate that the silhouette may or may not be made by that object. Then have students guess what object actually produces this silhouette. Explain that students are to draw a similar "mystery silhouette" of an object they find in their own home.

Note: If you can send bags of cubes home with students (26 cubes each), consider presenting the Excursion in the following session, Different Views of a City (p. 40), as homework rather than as an in-class session.

🔲 Extension

Left-Side, Back, and Bottom Silhouettes   As a challenging project, have students go back to any of the cube buildings on Student Sheets 10–13 and draw the silhouettes of three further views: the left side, back, and bottom. They can draw their silhouettes on one-centimeter graph paper. Tell them to label each drawing according to the view it shows.

Drawing the left side, back, and bottom silhouettes is more difficult because these views of the building cannot be seen in the pictures. Thus, most students will have to make the buildings with cubes first.

Ask students if they can find any relationship between the silhouettes of different views. As a hint, you might ask:

How are the front and back silhouettes related? How are the two side silhouettes related? What about the top and bottom silhouettes?

❖ **Tip for the Linguistically Diverse Classroom**   Encourage students to show any relationships visually; that is, they can cut out the silhouettes of their three further views, and place any cutout next to its related silhouette on the worksheet.

In each case, one view is the flip or mirror image of the other. Some students may not see this because when they change views, they lose track of the original orientation. You might suggest that they keep their building in one place on the table while they move around to view it from different

Staying Properly Oriented

positions.

Once students have made a cube building and are trying to draw its silhouettes, they may pick up their building and turn it without keeping track of which side is the front. Then they get confused. After pointing out this potential problem, you might suggest that they keep their building flat on the desktop or table while they move themselves around to view the building from different sides.

You could also have students use a small toy figure to act out viewing the building from different perspectives. Placing a cube building on a table, students should identify the front of the building and draw a "road" passing in front of it (or use a pencil to represent the road). They then place the toy figure looking directly at the front of the building and crouch behind the toy—with their eye level matching that of the toy—to view the building as the toy does.

Next, students make their toy figure "fly" straight up several feet, staying in front of the building, and look down. What does the top of the building look like from this front view? Students then return the toy to its original ground position and make it "walk" around to the right side of the building. They should again crouch to see the right side of the building as the toy figure would.

These activities should help students identify the different views of the cube buildings; a student who is having difficulty may reenact them at any time. However, even when students can correctly identify the different views, they still might not be able to properly visualize the corresponding silhouettes. For such students, always use the overhead to clarify what the building's actual silhouettes look like.

Integrating Three Views: How Students Try to Do It

When we ask students to construct an object by looking only at three silhouettes (as on Student Sheet 14), they must combine the information from the three different views and synthesize it to form one comprehensive view of the object. This integration process—being able to synthesize and combine different pieces of information into one comprehensive idea—is important not only in geometry, but in mathematics and reasoning in general.

As students work on these tasks, you will observe different levels of sophistication in their skills.

- Some students do not see at first that integration is necessary. They might, for example, make one building for each view.

- Other students recognize that integration is needed, but are unable to do it mentally. They look at one of the silhouettes and develop a conjecture about what the building will look like based solely on that one view. They then make a building to match that silhouette and physically check to see if it matches the other two views. If it doesn't, the students make successive changes to the building—taking away a cube here, adding one there—in an attempt to make it match all three views. As they work, they focus on only one view at a time.

- Some students also first make a building that matches one view, but they seem to expect that their building will require changes. They reflect on any changes before making them, trying to figure out how each change will affect the different views. They are often able to anticipate the effects of changes before they perform them. This approach, however, is one of coordination rather than integration of the three views.

- Finally, some students will begin to truly integrate the information given in the three silhouettes. They will approach the problem by mentally integrating the three views to make a mental model of the cube building—one that may be something like the isometric drawings they have seen throughout this unit. They then construct the building from their mental image.

Students who are having difficulty integrating the different views simply need more practice, more experience with this type of problem. However, don't feel that you must bring all students to the highest level of sophistication. For many students, this is a problem they have never before encountered, and they will need many such experiences over a span of several years before they are truly capable of mentally integrating different views into a single model.

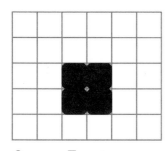

2.　　　Front

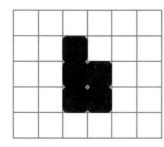

Top, as seen from the front

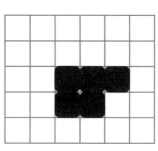

Right side

Describing Our Building Silhouettes

This discussion pertains to the silhouettes students have drawn for Building 2 on Student Sheet 11.

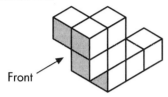

Front

Joey, tell me how you're visualizing the front silhouette of this figure.

Joey: Two squares going across. One coming down from the square on the right.

The teacher draws:

Did anybody draw a different silhouette?

Kumiko: Mine was just like Joey's, but had two coming down on the right.

The teacher draws:

We have two different versions of the front silhouette. Why?

Rebecca: Joey forgot to draw the back cube.

What do you think about that, Joey?

Joey: Kumiko is right. When I look at my building, it looks like hers.

OK. Let's check it out on the overhead. [*The teacher positions the cube building to confirm that Kumiko's idea for the front silhouette is correct.*] **Now let's try the top-view silhouette. What does it look like? How should I draw it?**

Luisa: Two cubes across and one on the top right.

Marci: Four in a straight line [*motions vertically*], and one coming out at the bottom on the left.

DeShane: Four across [*motions horizontally*], and one on top, at the left.

The teacher now has three different top views drawn.

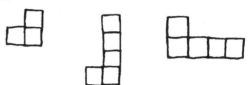

What do you think about these different answers?

Lesley Ann: I think that Luisa might not have drawn the back two cubes.

Will these two cubes show up in the silhouette?

Luisa: I don't know [*pause*] . . . Could we try it on the overhead?

The teacher places the cube building on the overhead so that the correct top silhouette projects onto the screen.

Luisa: Yeah, you can see them.

Qi Sun: Marci was right [*referring to the middle picture above*]. That's what I had, too.

Sarah: The drawing on the right is from the side.

Is it a top silhouette?

Sarah: Yes. It's really the same one, but turned around.

This discussion illustrates how students typically describe the silhouettes they draw. Some teachers encourage students to use more-precise language by drawing exactly what the students say, making intentional mistakes whenever their descriptions are vague. The dialogue also illustrates two common difficulties that students encounter as they attempt to draw silhouettes: disregarding recessed cubes, and using a side (rather than front) perspective when drawing the top view.

Different Views of a City

Materials

- Interlocking cubes (26 per student)
- Student Sheet 18 (1 per student)
- One-centimeter or three-quarter-inch graph paper (1 sheet per student, optional)

What Happens

Given a "map" of a cube city with the height of each building marked, students try to identify the locations from which different silhouettes of the city could be seen. Their work focuses on:

- understanding geometric perspective
- learning to visualize what objects look like from different perspectives

Where Was Each Photo Taken?

Hand out Student Sheet 18, Different Views of a City, which gives the top view of a cube city and the height of each building in the city. Read the directions together, then ask students to explain the problem in their own words. Make sure they understand what the numbers on the squares mean.

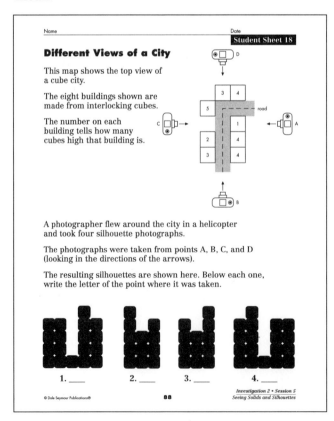

Can you predict the point from which each photograph was taken, before you build the city with cubes? If you can, write your predictions on your sheet.

Students then build the city and check their predictions. Students can build their city on graph paper. If they are using centimeter cubes, they can place the cubes directly on the activity sheet.

Some students will forget that cubes "in the back, on the other side of the street" will still appear in the silhouettes. Help them focus on this problem:

Would the buildings in back show up in the silhouettes? Why or why not?

These students should build the city with cubes first, then figure out what the silhouettes look like.

INVESTIGATION 3

"How-To" Instructions for Cube Buildings

What Happens

Session 1: Writing "How-To" Instructions
Students write instructions that tell others how to put together a simple cube building. They then watch another student try to use these instructions. On the basis of what happens, they evaluate the clarity of their instructions and revise them as needed.

Sessions 2 and 3: Which Instructions Are Best?
Students experiment with several different types of building instructions: 3-D pictures, three-view and two-view diagrams, layer-by-layer plans, and written directions. They then evaluate the effectiveness of these different types.

Mathematical Emphasis

- Interpreting different types of instructions for building with cubes
- Evaluating the effectiveness of different forms of "how-to" instructions
- Developing visualization skills
- Integrating information given in separate views or presented verbally to form one coherent mental model of a cube building

What to Plan Ahead of Time

Materials

- Interlocking cubes: 50–60 per pair (Sessions 1–3)

Other Preparation

- Duplicate student sheets and teaching resources (located at the end of this unit) in the following quantities. If you have Student Activity Booklets, no copying is needed.

For Session 1
Student Sheet 19, More Cube Buildings (p. 89): 1 per student (homework)

Student Sheet 16, Cube Silhouettes (p. 86): 1 per student (homework)

For Sessions 2–3
Student Sheet 20, Building Instructions Set A: 3-D Picture (p. 90): 1 per pair

Student Sheet 21, Building Instructions Set B: Three Straight-On Views (p. 91): 1 per pair

Student Sheet 22, Building Instructions Set C: Two Straight-On Views (p. 92): 1 per pair

Student Sheet 23, Building Instructions Set D: Layer-by-Layer Plans (p. 93): 1 per pair

Student Sheet 24, Building Instructions Set E: Written Directions (p. 94): 1 per pair

Student Sheet 25, Quick Image Geometric Designs (p. 95): 1 per student (homework)

- Make and cut apart a transparency of Quick Image Geometric Designs (p. 103) for the associated Ten-Minute Math activity.

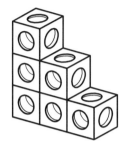

1. Take 6 cubes.
2. It looks like a staircase.
3. three on the bottom, 2 on top of the 3, the one on top of the 2.

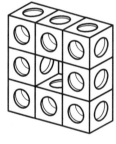

my building has 4 equal sides.
I used 8 blocks. It has a hole
the size of one block in the middle.
It's a 3×3 square.

make two straigt lines of 4.
Conect them on the top cube
with one cube. Conect them
on the botom cube with
one cube to make it look
like a rectangle.

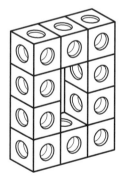

This building uses 10 cubes. First
make five cubes going up. On the
bottom put a cube going right.
On the next block up put a cube
to the left. On the next block put
one to the right. On the next block
put a cube to the left. On the top
block put a cube to the right.

Writing "How-To" Instructions

Materials

- Interlocking cubes (32 per pair)
- Student Sheet 16 (1 per student, homework)
- Student Sheet 19 (1 per student, homework)

What Happens

Students write instructions that direct others how to put together a simple cube building. They then watch another student try to use these instructions. Based on what they see, they evaluate the clarity of their instructions and revise them as needed. Their work focuses on:

- communicating about three-dimensional objects
- evaluating the effectiveness of their own communication
- developing visualization skills

 Ten-Minute Math: Quick Images Once or twice in the next few days, outside of the math hour, do the Quick Images activity. This time, use images cut from the Quick Image Geometric Designs transparency (included with the blackline masters at the end of this unit). Students will need pencil and paper.

Flash one design for 3 seconds; let students try to draw it.

Flash it for another 3 seconds, and let students revise their drawings; then leave the design visible for a final revision.

After finishing an image, students should talk about what they saw in successive flashes. You may hear comments like, "I saw four triangles in a row."

If students are having difficulty, have them try this strategy:

Each design is made from familiar geometric shapes. Find these shapes and try to figure out how they are put together.

As students describe their figures, you can begin to introduce correct terminology for these shapes—parallelograms, octagons, and so forth.

For variations and full directions for this activity, see p. 63.

Writing Good Instructions

In this unit we've been making buildings by putting together cubes. How have you known where to put the cubes? Did the pictures give you all the information you needed to put together each building? Are there other kinds of instructions that tell us how to build things? What kinds of instructions have you seen or used for building toys or models?

Briefly explain the activity: Each student makes a simple cube building out of 6 to 8 cubes, then writes or draws instructions that will tell somebody else how to build the same building.

Your job is to make a clear set of instructions that will help someone else make exactly the same thing you made. Your instructions can use words or drawings or both. Keep in mind that the person you are writing these for has never seen your building.

When everyone has completed a building and a set of instructions, the students hide their buildings and trade instructions with a partner. (Ideally, try to pair students who haven't been seated together.) Students then use the instructions they are given to construct the intended building. When they are satisfied, they compare their building with the one the writer of the instructions intended (which has been hidden until now). Together, they discuss the possible cause for discrepancies. At this point, students can add to or change their directions to make them clearer.

As a class, discuss which type of instructions were most useful and which were more difficult to follow. Ask students how they would change their instructions if they could make a new set.

Session 1 Follow-Up

More Cube Buildings and Cube Silhouettes   On Student Sheet 16, Cube Silhouettes, students draw the front, top, and right-side silhouettes of the three buildings that appear on Student Sheet 19, More Cube Buildings.

 **Homework**

Which Instructions Are Best?

Materials

- Student Sheets 20–24 (1 each per pair)
- Student Sheet 25 (1 per student, homework)
- Interlocking cubes (50–60 per pair)

What Happens

Students experiment with several different types of building instructions: 3-D pictures, three-view and two-view diagrams, layer-by-layer plans, and written directions. They then evaluate the effectiveness of these different types. Their work focuses on:

- interpreting different types of instructions for building with cubes
- evaluating the effectiveness of different forms of "how-to" instructions
- developing visualization skills
- integrating information given in separate views or presented verbally to form one coherent mental model

 Ten-Minute Math: Quick Images In short sessions at various points in the school day, continue to do the Quick Images activity with geometric designs, as described in Session 1.

For variations and full directions for this activity, see p. 63.

Activity

Trying Out Different Instructions

Pair students for this activity and give each pair one copy of Student Sheets 20–24, Building Instructions Set A through Set E, the five different types of directions for making 10-cube buildings.

Let's say you work for a toy company that sells little building cubes for children. Your boss has asked you what kind of directions the company should provide to tell children how to make different cube buildings.

These five sheets are examples of five different kinds of directions that the company is thinking about using. Your job is to figure out how well each set works.

Student partners work together to try making the specified ten-cube building from each set of instructions.

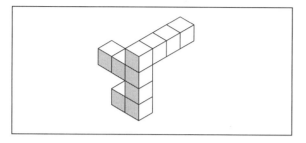

Student Sheet 20: 3-D Picture

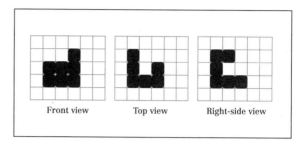

Student Sheet 21: Three Straight-On Views

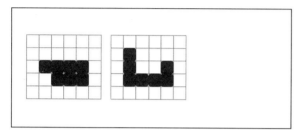

Student Sheet 22: Two Straight-On Views

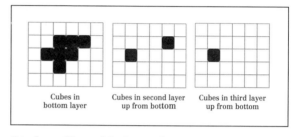

Student Sheet 23: Layer-by-Layer Plans

Take five cubes and attach them to make a straight line. Take another three cubes and attach them to make a straight line. Lay both lines flat on a table, with both horizontal. Place the shorter line of cubes behind the longer line so that the shorter is placed evenly between the ends of the longer. Attach these two lines together.

Now connect the two remaining cubes. Hold these two cubes so that they are sticking up from the table—they should be vertical. Now attach these two cubes to the top of the last cube on the right of the shorter line of cubes.

Student Sheet 24: Written Directions

As they work, they are to critique the different types of instructions, telling what is good and bad about each and how they could be improved. Have students keep each cube building, with its instructions, so that they can be examined and compared in the class discussion.

∗ **Tip for the Linguistically Diverse Classroom** If possible, ask a parent, aide, or teacher to translate the written directions on Student Sheet 24 into students' native languages. If this is not possible, or if students cannot read in their first language, read (or have a fluent English-speaking partner read) the directions orally in English. After each instruction, have the students actually do what was directed. Without this modification, limited-English-proficient students will not be able to offer their own ideas about how the cubes should go together, and whether the directions are good ones.

When everyone has worked with all five instruction sets, have the class discuss its findings:

Were you able to make each cube building? Everyone hold up the buildings you made from instruction set A, the 3-D picture. Are your buildings all the same? How are they different? What was good and bad about this set of instructions?

Repeat questions like these for each instruction set. Read the **Teacher Note**, Students' Thinking About Instructions (p. 49), for typical student evaluations of the different types.

Which instructions were best? Why? Which were the worst? Why? How did you decide which instructions were best and worst? Would most children be able to understand these instructions?

❖ **Tip for the Linguistically Diverse Classroom** For nonverbal responses, students can use the student sheet numbers (20–24) to rank the different types of instructions from best to worst, or to list them on a chart under the columns "good" (thumbs up) and "bad" (thumbs down), according to what they decide.

What makes a "good" set of instructions? That is, are instructions "good" if the building is easy to make? What if the building is easy to make, but everyone made a *different* building—would those instructions be good? Why or why not?

What if a set of instructions said simply, "Make a building from 10 cubes." Would that building be easy to make? Would we all make the same building? Why or why not? Would this be a good set of instructions?

Try to get students to think about the idea that good instructions are both easy to follow and clear enough so that everybody who reads them makes the intended building.

Activity

Sessions 2 and 3 Follow-Up

 **Homework**

Quick Image Geometric Designs Students play Quick Images with a family member, using the geometric designs on Student Sheet 25, Quick Image Geometric Designs.

Students' Thinking About Instructions

In their evaluations of the different sets of building instructions on Student Sheets 20–24, most students try to take the point of view of others, saying things like "This [set] is OK for young children," or "This is too hard for little kids—for big kids it's fine." You might point out as well that different kinds of instructions may be preferred by—and even work better for—people with different strengths and different styles of thinking. Following are some typical student reactions.

Building Instructions Set A: 3-D Picture This form of instruction seems to be the clearest for many students. Typical remarks:

> There are no words. So a little kid can read it.

> You can see how they are connected.

> The shading helps to tell you the sides.

Building Instructions Set B: Three Straight-On Views This is generally the most difficult set of instructions, as discussed in the **Teacher Note**, Integrating Three Views: How Students Try to Do It (p. 38). Remarks:

> [This set is good because] you know that this is the only way that this can possibly be.

> These were hard. I couldn't figure out what it was.

Building Instructions Set C: Two Straight-On Views Some students believe these instructions are difficult to follow because of their ambiguity. The instructions do not say which views—front, top, right side, back—are being shown. Other students like this set better than the three views of set B because they think these are "easier"— meaning that it is easier to find a building that has these two views. In fact, it is easier to find a building that has two given views rather than three because there are fewer constraints to satisfy. There is more flexibility with two views— students can decide how to interpret the drawings. One student pair reacted this way:

> These are going to be terrible. They're the same as the three views.

> No, wait. These aren't labeled and there's only two. So you can build it any way you want.

As precise building instructions, two views will usually not be as good as the standard three views given in set B. The flexibility impedes communication—they describe more than one building.

Building Instructions Set D: Layer-by-Layer Plan Most students understand this set of instructions intuitively, saying things like this:

> The bottom is easy.

> Layers. They're like in a cake. Or like the stories in a building.

Even so, some students encounter difficulties. For instance, some do not properly coordinate the relative positions of the cubes as they move from layer to layer. They may not know where to attach the second-layer cubes to the bottom layer.

Building Instructions Set E: Written Directions Some students like the verbal description because it explicitly tells them what to do, step by step. However, interpreting the intended meanings of the words and creating mental images to guide the construction can be very difficult. Some student remarks:

> I like the way these tell you exactly what to do.

> These are awful. I don't understand what this part means.

The Cube Toy Project

What Happens

Sessions 1, 2, 3, and 4: Making Plans for a Cube Toy Students create their own toys out of 50–60 interlocking cubes. They then write instruction booklets that will enable a younger child to build their toys. As they put together these booklets, they choose and combine the types of instructions that will best communicate their plans. They then evaluate the effectiveness of their instructions both by using their booklets themselves and by asking other students to try building the toys. They pass through several rounds of revision as they clarify their instructions.

Mathematical Emphasis

- Interpreting different types of instructions for building with cubes
- Evaluating the effectiveness of different forms of "how-to" instructions
- Developing visualization skills
- Communicating effectively about three-dimensional objects

What to Plan Ahead of Time

Materials

- Interlocking cubes: 120 per student (or per pair, if cubes are in short supply) (Sessions 1–4)

Other Preparation

- Duplicate student sheets and teaching resources (located at the end of this unit) in the following quantities. If you have Student Activity Booklets, no copying is needed.

Sessions 1–4

One-centimeter or three-quarter-inch graph paper, (pp. 100 and 101; choose the size that more closely matches the size of your cubes): 1–8 sheets per student (class), 2 sheets per student (homework)

Isometric grid paper (p. 99): 1–2 sheets per student

Student Sheet 26, Quick Image Dot Patterns (p. 96): 1 per student (homework)

Student Sheet 27, Make a Quick Image (p. 97): 1 per student (homework)

- Make a transparency or a laminated display sheet from the blackline master Isometric Grid Drawing (p. 98).

- Make and cut apart a transparency of Quick Image Dot Patterns (p. 104) for the associated Ten-Minute Math activity.

Making Plans for a Cube Toy

Materials

- Interlocking cubes (120 per student)

- One-centimeter or three-quarter inch graph paper (1–8 sheets per student, class; 2 sheets per student, homework)

- Isometric grid paper (1–2 sheets per student)

- Isometric Grid Drawing (transparency or laminated sheet)

- Student Sheet 26 (1 per student, homework)

- Student Sheet 27 (1 per student, homework)

What Happens

Each student makes a toy from interlocking cubes and an instruction booklet that tells others how to make the toy. **Note:** There are only 60 cubes per student in your kit. Thus to present the activity as written, you will need to borrow cubes from another teacher for just this session. If that is not possible, you can arrange for students to do this activity in pairs, with each student still writing and revising her or his own instructions for the toy. Students' work focuses on:

- interpreting different types of instructions for building with cubes
- evaluating the effectiveness of different forms of "how-to" instructions
- developing visualization skills
- communicating effectively about three-dimensional objects

 Ten-Minute Math: Quick Images   Once or twice during the next few days, when you have a spare 10 minutes, do the Quick Images activity, this time with dot patterns. Use patterns cut from the Quick Image Dot Patterns transparency. Students will need pencil and paper.

Tell students that when you flash a pattern, they must think about two questions:

> **How can you draw the dot patterns you saw?**
> **How can you figure out how many dots you saw?**

Flash one pattern for 3 seconds. After students try to draw it, flash it for another 3 seconds, and let students revise their drawings; then leave the pattern visible for a final revision.

Ask how many dots they saw in the pattern. You may notice students using different strategies: Some will see a multiplication problem (such as 6×3) and will not draw the dots until reminded to do so; others will draw the dots, then figure out how many there are.

For variations and full directions for this activity, see p. 63.

Building the Toy

Let's imagine that you're still working for the toy company that makes and sells building cubes. Your next project is to design and build a new toy out of cubes. Then you will write an instruction booklet that tells second or third graders how to make that toy themselves.

Have each student build a "toy" out of 50–60 cubes. The 60-cube limit is essential. (They will need the remaining cubes for trying to build other students' toys.) If the limit is exceeded, students spend too much time building the toys, and their plans become too complicated. We suggest that you specify nonviolent toys only—no weapons; this will encourage some students to be more creative in their choice of toys.

Students must save their toys, perhaps in small grocery bags. Allow approximately 30–40 minutes for designing and building the toys. After this time, students should not be allowed to change their designs.

Although students will be quite excited about making the toys—it allows them to be creative—the real value of this project comes from writing and drawing the assembly plans, as explained in the following activities.

Making the Instruction Booklet

Students prepare a draft of an instruction booklet that will enable a second or third grader to build their toy. We call it a draft at this stage because students should expect to change and improve it. You can also relate the steps that the students take in writing their booklets to the process they follow when they write papers in general: first draft, peer review, revision, and so forth. At draft stage, they should work in pencil.

We have talked about different ways you can describe how to make cube buildings. You can decide which of these to use for your instruction booklet. Just remember, an instruction booklet is good if the person reading it can easily and correctly make the object that is described.

Review with students the different methods used to describe how to build cube figures: written instructions; 3-D pictures; drawings of top, side, and front straight-on views; layer-by-layer plans; drawings of two straight-on views; or any combination of these. When students want to include written directions, remind them that younger children might not read as well as they do in fourth grade.

You might suggest that students try giving the instructions in parts: first, instructions for building one part of the toy, then for another part, and so on. This strategy is used very effectively in instruction booklets provided with many Lego building sets or models.

If students choose to draw separate views or layer-by-layer plans, give them one-centimeter or three-quarter-inch graph paper. If students choose to draw 3-D pictures, provide them with isometric grid paper. Use the blackline master Isometric Grid Drawing—making either a transparency or a laminated sheet—to illustrate how this paper can be used.

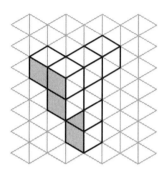

Initial Evaluation and Revision   After all students have completed a first draft of their instructions, have them store their toys out of sight. Then, students should try to build a second copy of their toys with the remaining 60 cubes, relying on their own instructions. Suggest that they try to pretend not to know what their toy looks like—this will help them evaluate whether their directions tell clearly how to build it.

Would somebody who has never seen your toy be able to understand your instructions?

Circulate around the room and help students evaluate their plans. Give students time to make revisions.

<hr>

Activity

Testing the Plans

Have students trade their instructions with a partner. Each student will use the plans to build the toy described. Advise students not to give any oral instructions to their partners during this building time. Also, all original toys should be kept out of sight.

As students are building, they write suggestions to the author of the plans on how to improve them. When they feel they are done, the students compare the toy they built to the original. Student partners then discuss how the instructions could be changed to make them easier to follow, and to reduce the likelihood of mistakes.

❖ **Tip for the Linguistically Diverse Classroom** Rather than writing, students might show or draw their suggestions for improving the directions.

After all the students have had their draft instructions tested by a classmate, conduct a whole-class discussion in which students discuss the strategies they used in their instructions and the difficulties that they encountered. They discuss how their plans could be improved. Each student then makes a final version, incorporating the suggested revisions.

Finally, invite a class of second or third graders to use the booklets to build the toys. Each author observes the use of his or her booklet. Afterwards, have a class discussion about how well the plans worked for the younger students.

Assessment

Evaluating the Instruction Booklets

You can assess students' learning during this unit by examining their final building instructions. After you collect students' toys and instruction booklets, evaluate the clarity of each set of instructions: Is this a reasonably clear explanation for building the toy? Note, however, that students will make toys of varying complexity. Telling how to build some toys may be relatively easy, while others are quite difficult to describe.

Also, examine students' use of the different methods for representing 3-D objects with drawings: Which methods did students choose to use in their instruction booklets? Did they use them effectively? See the **Teacher Note**, Evaluating Students' Instruction Booklets (p. 57).

You will likely find several levels of sophistication in students' building instructions:

- Some students will have difficulty writing clear instructions because they fail to take into account the reader's perspective. That is, their instructions may be clear to them, but not to a person who does not already know what they are talking about.

- Other students will write very clear instructions for building rather simple toys, but will not employ any of the drawing techniques discussed in this unit. For example, they might draw one straight-on front view of their toy.

- Finally, some students will make a complex toy and employ several of the drawing methods discussed in class to effectively describe the construction of their toy.

Choosing Student Work to Save

As the unit ends, you may want to use one of the following options for creating a record of students' work on this unit.

- Students look back through their folders or notebooks and write about what they learned in this unit, what they remember most, what was hard or easy for them. You might have students do this work during their writing time.

- Students select one or two pieces of their work as their best work, and you also choose one or two pieces of their work, to be saved in a portfolio for the year. You might include the halves of Student Sheet 13 from the assessment, Drawing Before Building (Investigation 2, Session 3), and any other assessment tasks from this unit. Students can create a separate page with brief comments describing each piece of work.

- You may want to send a selection of work home for parents to see. Students write a cover letter, describing their work in this unit. This work should be returned if you are keeping a year-long portfolio of mathematics work for each student.

Sessions 1, 2, 3, and 4 Follow-Up

 **Homework**

Quick Image Dot Patterns   After Session 2, students take home Student Sheet 26, Quick Image Dot Patterns, to play with someone at home.

Make a Quick Image   After Session 3, students make their own Quick Image Dot Pattern by arranging 20 dots on a sheet of graph paper in a way they think is easy to remember. They might test it out on someone at home to see if another person can copy their design. You might talk with students about what is a reasonable degree of challenge; if their arrangements are too complex, people will just become frustrated.

Each student will need two sheets of blank graph paper and a copy of Student Sheet 27. You might also remind students of the examples of dot patterns they already took home, on Student Sheet 26.

Remind students to bring their Dot Patterns back to class. Students can trade the images with a partner, or you might make transparencies of their examples and use them for Ten-Minute Math: Quick Images.

 **Extensions**

Talent on Display   The toys and instruction booklets make an impressive display of the students' creativity and visual/graphics skills (say, for an open house).

Wordless Directions   Challenge students to make an instruction booklet that uses no words. (You might show booklets for Lego building blocks as an example.) Discuss why it might be good to make instructions with no words (for example, so they can be used by people who cannot read or who speak a different language).

Evaluating Students' Instruction Booklets

Making the instruction booklets for the cube toy project is a challenging task. It is difficult for students both to write clear instructions for building a toy and to coordinate these instructions with appropriate drawings.

Several factors affect the difficulty of this task. Some students build and write instructions for "flat" toys of only one layer. Others build much more complicated toys that necessitate multiple drawings and a complex set of written instructions. Some toys have a recognizable shape (for example, pig, house, robot), and some are more abstract. Assessment of students' instructions should take into account this great variability in difficulty.

The following examples show the range of instructions that students may produce. In reading these examples, keep in mind the following criteria for assessing students' work: clarity, consistency (is one part of the instructions consistent with other parts?), completeness, use of appropriate methods for describing how the toy should be built, evidence of careful proofreading for mistakes.

The first three students presented their written instructions and their drawings separately. The last two student pairs integrated theirs. Note that Nick (the radio) and Jesse (the pig) did not use the required 50–60 cubes.

The Radio

Nick's written instructions (below) are incomplete, at first seem to have no sequence, and are thus unclear. He forgot to label his front view, but examining it while reading his written instructions gives a pretty good idea of what he intended. Nick's other views are correct, but he drew the right-side view twice (or perhaps didn't erase an earlier version properly), making that view confusing. Nick also used X's to signify individual cubes; this method is a meaningful but nonstandard way to represent cubes.

The Radio

First make seven going across in a solid color. Now make two cubes in a different color, then three cubes in the original color, then two cubes in the other color. On the next row do the exact same thing. Now make seven going across in the original color. On the next row put one cube on the first cube and one on the seventh cube.

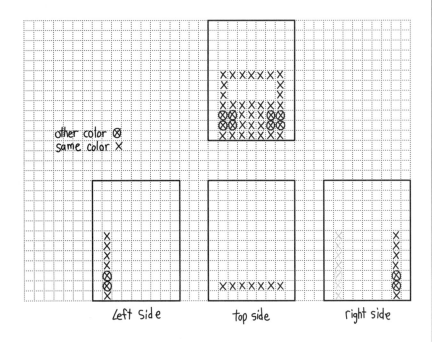

Continued on next page

The Pig

Jesse gave a page of written instructions for making a pig, accompanied by four more pages of drawings showing each of the pig's layers. While there are a couple of ambiguities (how should the three rows "connect"?), the instructions are for the most part clear and direct.

Further, the drawings of each layer, which Jesse carefully superimposed over each other on consecutive pages of his booklet, show precisely how the structure fits together. However, he has, like Nick, used a nonstandard method of representing cubes—X's rather than squares.

My pig uses 31 cubes. First you make two rows of four cubes and then one rows of five and put them horizontal and put the long one in the middle and connect them.
Then you put one cube on the bottom of each corner. Then you make three rows of four cubes and put them horizontal and connect them and then put it on top of the other three rows of cubes. Then take two cubes and put them on the top front corners.

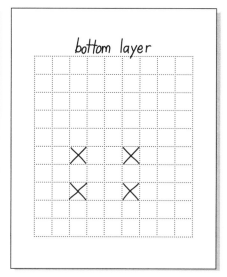

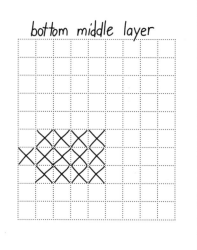

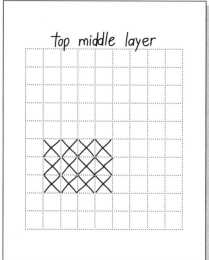

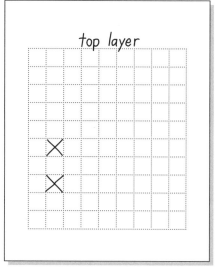

Continued on next page

The House Jacuzzi

Lina Li's toy resembled a rectangular prism. She decided that the most straightforward instructions would show the reader how to build its rectangular faces, which she described step by step. Sometimes she forgot, however, to specify exactly what she intended. For example, her description of the roof did not explain that the roof must be rotated about a horizontal line before it is inserted between the sides. But Lina Li supplemented her work with several straight-

on drawings, which help clarify some of the ambiguities in her verbal instructions. Thus, Lina Li has broken down a potentially complex problem into component parts that are, as a whole, sufficient to correctly build her toy. Even though her front view is actually a front *layer* rather than a straight-on view, Lina Li's use of straight-on views is overall very effective in communicating her ideas.

First make a complete square with 4 down and 4 across. Repeat this process for the left wall and stand them both up. Now make another 4 by 4 square and insert it like this for the roof.

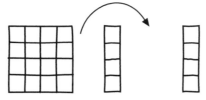

Snap to first cube on inside, now you've made the top. Now to make the back wall make a complete rectangle with 4 across and 3 down.

For the front attach 3 cubes on the front inside going down the right and left sides. Next attach 2 cubes to the inside top and you're done.

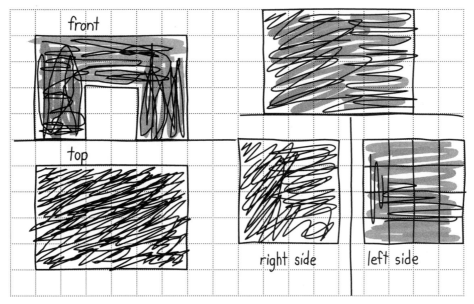

Continued on next page

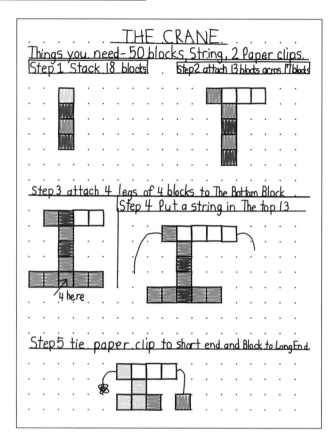

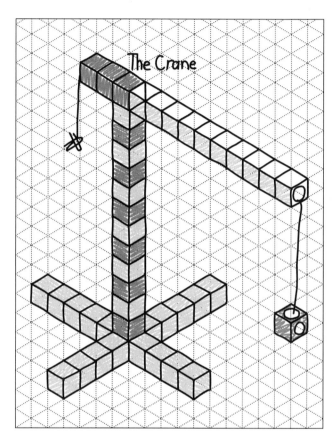

The Crane

Kyle and Ahmad have effectively interwoven written instructions with straight-on views of their toy in describing how it should be built, step by step. Even so, their toy would be difficult to build if a child had only these instructions. To help the builder, they added a very nice isometric drawing of their crane. The only weakness in their work is some inconsistency between their isometric drawing and their step-by-step instructions. For example, in Step 3, they direct the builder to connect four legs of 4 cubes to the bottom of the crane. But in the isometric picture, they drew two legs with 4 cubes, and two with 5 cubes. Kyle and Ahmad needed to more carefully proofread their instructions. Overall, though, we see that these students have used great imagination in designing their toy, and have used several methods of communicating how to build it.

Continued on next page

Balancing Boy

Sarah and Irena have effectively interwoven detailed written instructions with diagrams. Because their toy has only one layer, they used only straight-on views. They have done an excellent job of segmenting their instructions into easy steps, with one set of instructions for the boy, and another set for the balance beam. They divided the figure into meaningful parts, which they named. They used color effectively, and they told exactly how many cubes of each color were needed for each part. Finally, Sarah and Irena drew a single picture of the finished toy to show how the two major parts fit together. Sarah and Irena's written instructions were so detailed that a builder could probably make the toy successfully, even without looking at the final picture.

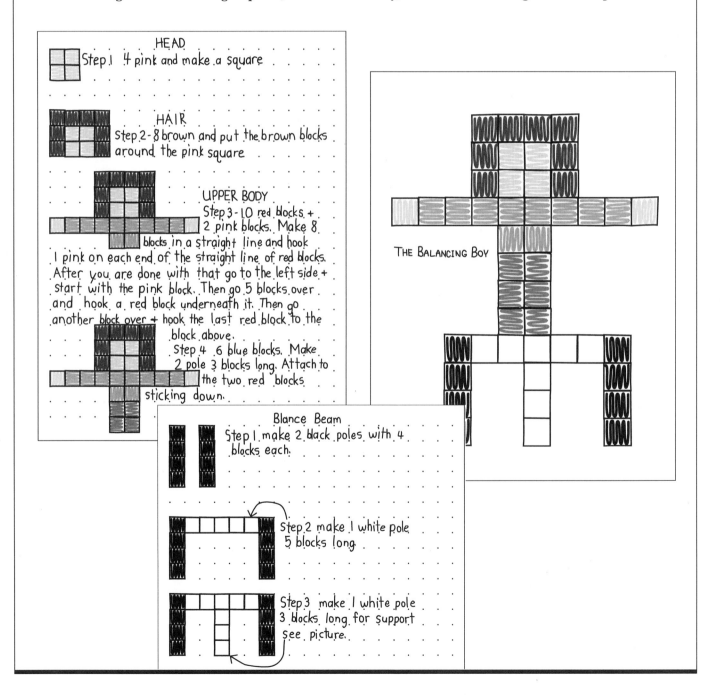

HEAD
Step 1 4 pink and make a square

HAIR
Step 2 - 8 brown and put the brown blocks around the pink square

UPPER BODY
Step 3 - 10 red blocks + 2 pink blocks. Make 8 blocks in a straight line and hook 1 pink on each end of the straight line of red blocks. After you are done with that go to the left side + start with the pink block. Then go 5 blocks over and hook a red block underneath it. Then go another block over + hook the last red block to the block above.
Step 4 6 blue blocks. Make 2 pole 3 blocks long. Attach to the two red blocks sticking down.

Blance Beam
Step 1 make 2 black poles with 4 blocks each.

Step 2 make 1 white pole 5 blocks long.

Step 3 make 1 white pole 3 blocks long for support see picture.

THE BALANCING BOY

Quick Images

Basic Activity

Students are briefly shown a picture of a geometric object. Depending on the kind of object, they either build it or draw it by developing and inspecting a mental image of it.

For each type of problem—cube images, 2-D shapes, and dot patterns—students must find meaningful ways to see and develop a mental image of the object. They might see it as a whole ("it looks like a box, three cubes long and two cubes high"), or decompose it into memorable parts ("it looks like four triangles—right side up, then upside down, then right side up, then upside down"), or use their knowledge of number relationships to remember a pattern ("there were 4 groups of 5 dots, so it's 20"). Their work focuses on:

■ organizing and analyzing visual images

■ developing concepts and language needed to reflect on and communicate about spatial relationships

■ using geometric vocabulary to describe shapes and patterns

■ using number relationships to describe patterns

Materials

■ Overhead projector

■ Overhead transparencies of the geometric figures you will use as images for the session; we provide a few pages of transparency masters to get you started. To use the images, first make a transparency, then cut out the separate figures and keep them in an envelope. Include the numbers beside the figures; they will help you properly orient the figures on the overhead.

■ Interlocking cubes: 20 per student (if you are using figures of cube buildings)

■ Pencil and paper (if you are using 2-D images or dot patterns)

Procedure

Step 1. Flash an image for 3 seconds. You might show a picture of a cube building, a geometric drawing, or a dot pattern.

It's important to keep the picture up for as close to 3 seconds as possible. If you show the picture too long, students will build from the picture rather than their image of it; if you show it too briefly, they will not have time to form a mental image. Suggest to students that they study the figure carefully while it is visible, then try to build or draw it from their mental image.

Step 2. Students draw or build what they saw. Give students a few minutes with the cubes or their pencil and paper to try to construct or draw a figure based on the mental image they have formed. After you see that most students' activity has stopped, go on to step 3.

Step 3. Flash the image again, for revision. After showing the image for another 3 seconds, students revise their building or drawing, based on this second view.

It is essential to provide enough time between the first and second flashes for most students to complete their attempts at building or drawing. While they may not have completed their figure, they should have done all they can until they see the picture on the screen again.

Step 4. Show the image a final time. When student activity subsides again, show the picture a third time. This time leave it visible, so that all students can complete or revise their solutions.

Step 5. Students describe how they saw the drawing as they looked at it on successive "flashes."

Variations

We provide transparency masters for three types of Quick Images: cube buildings, geometric designs, and dot patterns. You can supplement any of these with your own examples or make up other types.

Continued on next page

Quick Image Cubes   Each student should have a supply of 20 cubes and be seated facing the overhead screen. Show a picture from the Quick Image Cubes transparency; note that the transparency should be cut apart so that only one picture shows at a time. Proceed as described. You may be able to show two or three Quick Images in a 10-minute session. Some teachers have found it valuable to repeat, on successive days, Quick Images that were previously presented.

Quick Image Geometric Designs   Use the Quick Image Geometric Designs transparency. Follow the same procedure as for the cubes, but have students draw the images they see.

When students talk about what they saw in successive flashes, many students will say things like "I saw four triangles in a row." You might suggest this strategy for students having difficulty: "Each design is made from familiar geometric shapes. Find these shapes and try to figure out how they are put together."

As students describe their figures, you can introduce correct terms for them. As you use them naturally as part of the discussion, students will begin to use and recognize these geometric terms.

Quick Image Dot Patterns   Use the Quick Image Dot Patterns transparency. The procedure is the same, except that now students are asked two questions: "Can you draw the dot patterns you see? Can you figure out how many dots you saw?"

When students answer only one question, ask them the other again. You will see different students using different strategies. For instance, some will see a multiplication problem, 6×3, and will not draw the dots unless asked. Others will draw the dots, then figure out how many there are.

Using the Calculator   You can integrate the calculator into the Quick Image Dot Patterns activity. As you draw larger or more-complex dot patterns, students may begin to count the groups and the number of groups. They should use a variety of strategies to find the total number of dots, including mental calculation and the calculator.

Creating Quick Images   Students can make up their own Quick Images to challenge the rest of the class. Talk with students about keeping these reasonable—challenging, but not overwhelming. If they are too complex and difficult, other students will just become frustrated.

Paper Quick Images   Instead of using the overhead projector, students in a small group can simply show a paper picture for a few seconds; cover it up while other members of the group try to draw it; then show it again; and so forth. Members of the group take turns creating images for others in the group to try.

The following activities will help ensure that this unit is comprehensible to students who are acquiring English as a second language. The suggested approach is based on *The Natural Approach: Language Acquisition in the Classroom* by Stephen D. Krashen and Tracy D. Terrell (Alemany Press, 1983). The intent is for second-language learners to acquire new vocabulary in an active, meaningful context.

Note that *acquiring* a word is different from *learning* a word. Depending on their level of proficiency, students may be able to comprehend a word upon hearing it during an investigation, without being able to say it. Other students may be able to use the word orally, but not read or write it. The goal is to help students naturally acquire targeted vocabulary at their present level of proficiency.

We suggest using these activities just before the related investigations. The activities can also be led by English-proficient students.

Investigation 1

cube building

1. As students watch, connect about eight cubes to make a structure of some sort. (It need not resemble a real building.) Explain that you are *building* with cubes and that what you are making can be called a *cube building*.

2. Make two or three more different cube configurations, explaining that *cube buildings* can have many different shapes.

3. Challenge students to demonstrate comprehension of this term by giving them each about eight cubes and telling them to make a cube building. Have everyone display the finished buildings.

   Who has a cube building that looks like mine?

   Do you see a cube building made the same way as Luisa's?

   Who else has a cube building that is the same as someone else's?

Investigation 2

front, top, side, back

1. Display some small object (a doll, a toy car, an alarm clock) that has a clear front, top, side, and back. Walk around or turn the object as you identify the different views, motioning with a finger from your eyes to the appropriate part.

   If I look at it this way, I see the *top*.

   If I look at it this way, I see the *front*.

   From here, I am looking at the *side*.

2. Make a small, nonsymmetrical cube building. Place it next to the object you have been viewing, and relate the same words (*front, top, side, back*) to this cube structure.

3. Challenge students to demonstrate comprehension of these words by following action commands.

   Stand up when I point to the back of the [clock].

   Clap your hands when I point to a side of the building.

Investigation 3

instructions, directions

1. Show and identify instruction booklets from several games, toys, or models. Make sure each booklet is placed on or next to the corresponding product.

2. Open the booklets and point out that the written text and diagrams are called *instructions* or *directions*.

3. Challenge students to demonstrate comprehension of these words by following action commands. Interchange the words *directions* and *instructions*.

   Pick up the instruction booklet for the model airplane.

   Point to the direction booklet for these building blocks.

   Point to the first instruction in this booklet.

Continued on next page

layer

1. Use cubes to make a building with three layers. Point to and identify each layer.

   This is the *bottom layer* of the building.

   This is the *middle layer*.

   This is the *top layer*.

2. Challenge students to demonstrate comprehension by following action commands.

   Make a building with four layers.

   Show me the top layer.

   Make a building with more cubes in the bottom layer than in the top layer.

Blackline Masters

_____ , 19 ____

Dear Family,

For the next couple of weeks, our class will be doing a mathematics unit in 3-D geometry called *Seeing Solids and Silhouettes*. What your child will learn has very practical uses in daily life. When you read a floor plan, use a diagram to make a bookshelf, or sew a shirt from a pattern, you have to translate the words and drawings on a flat piece of paper into the actual object. This kind of "3-D thinking" also helps solve many problems in mathematics, engineering, and science.

Your child will be doing a lot of work with cube buildings—closely examining pictures of them and then making the buildings with cubes. Your child will also work on writing clear instructions for making things with cubes.

We'll be talking about silhouettes, too. We'll look at shadows made by real objects, and try to match them with the objects. And we'll talk about different perspectives—that is, how things look from different angles or different views.

If your child has any interlocking blocks, spend some time making simple buildings together. Look at what you build and try drawing them from different perspectives. If you're not very good at this, your child may be able to help, using strategies learned in class!

If your child builds models of any sort, talk together about the instructions that come with them. Are the words and diagrams clear? How could they be better? Any time that you're putting together a bicycle, a toy, or a piece of furniture, have your child help read and interpret the directions.

And if your child has never made something from a set of plans, encourage it— whether your child is a girl or a boy. Many people believe that either you're good at this or you're not. In fact, visualizing is a skill like anything else—the more you practice, the better you get!

Sincerely,

Make the Buildings

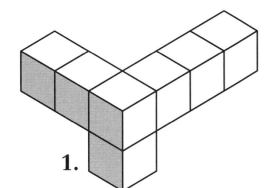

1.

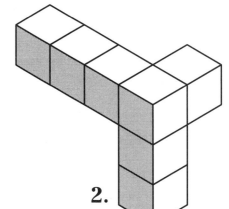

2.

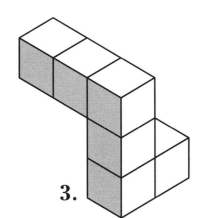

3.

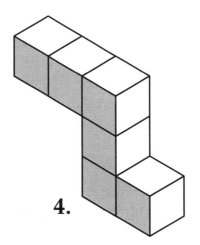

4.

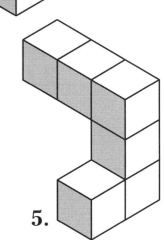

5.

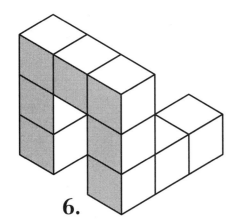

6.

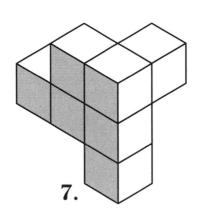

7.

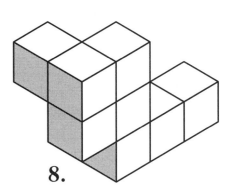

8.

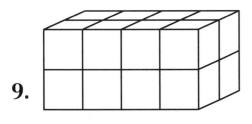

9.

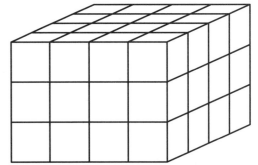

10.

How Many Cubes?

How many cubes does it take to make each building?
Predict. Then build with cubes to check.

1.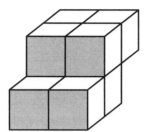

Prediction: _____ cubes

2.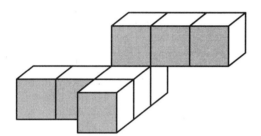

Prediction: _____ cubes

3.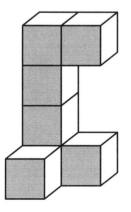

Prediction: _____ cubes

3-D Challenges

Challenge 1

Tell, without making it, how many cubes it would take to make this building?

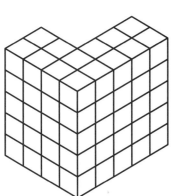

Challenge 2

Make this building with cubes. How many cubes did you use?

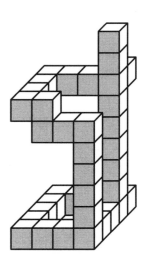

Challenge 3

Can you make a cube building like this?

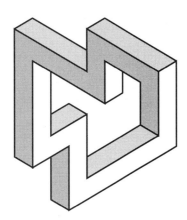

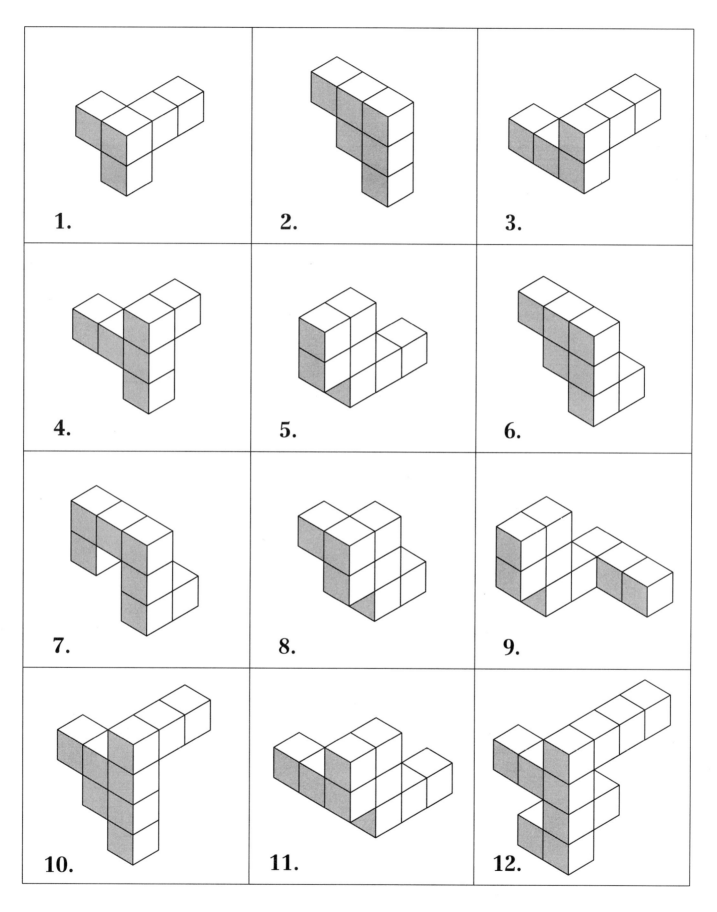

1.

2.

3.

4.

5.

6.

7.

8.

9.

10.

11.

12.

Investigation 1 • Resource
Seeing Solids and Silhouettes

Silhouettes of Geometric Solids

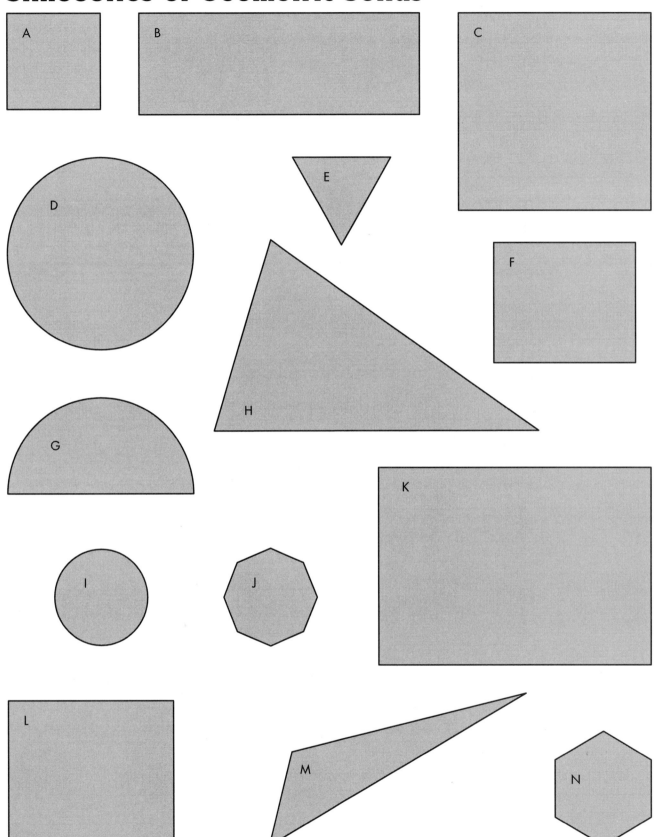

Landscape 1

Build this landscape with your geometric solids. Then look at each pair of silhouettes below. Find all the points in the landscape from which you could see both silhouettes in a pair. Write the letters of these points beside the silhouettes.

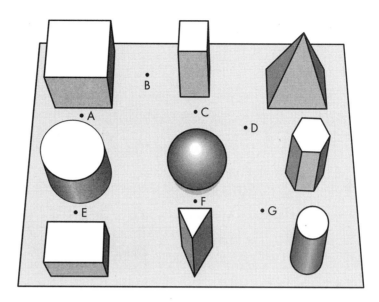

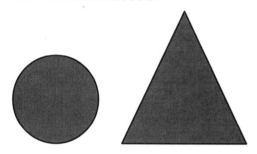

Pair 1

Points from which these could be seen:

_____

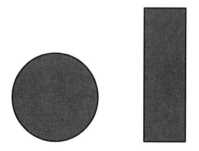

Pair 2

Points from which these could be seen:

_____

Pair 3

Points from which these could be seen:

_____

Landscape 2

Build this landscape with your geometric solids. Then look at each pair of silhouettes below. Find all the points in the landscape from which you could see both silhouettes in a pair. Write the letters of these points beside the silhouettes.

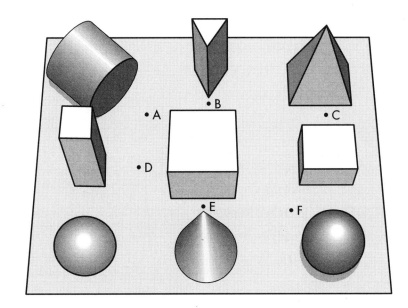

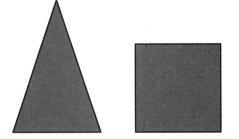

Pair 1

Points from which these could be seen:

_____

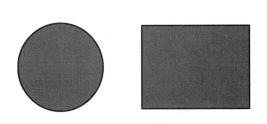

Pair 2

Points from which these could be seen:

_____

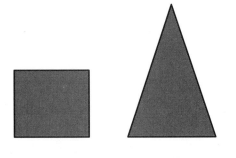

Pair 3

Points from which these could be seen:

_____

Landscape 3

Build this landscape with your geometric solids. Then look at each pair of silhouettes below. Find all the points in the landscape from which you could see both silhouettes in a pair. Write the letters of these points beside the silhouettes.

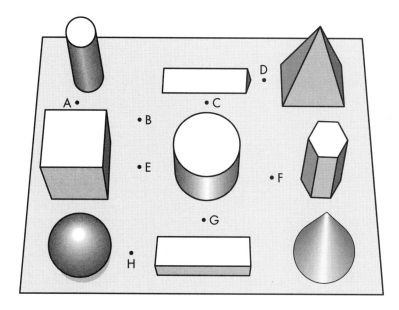

Pair 1

Points from which these could be seen:

_____

Pair 2

Points from which these could be seen:

_____

Pair 3

Points from which these could be seen:

_____

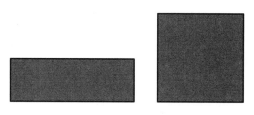

Landscapes Challenge

Look at this pair of silhouettes. Find all the points in the three geometric landscapes from which these two silhouettes could be seen. Write the letters of those points.

These could be seen from the following points:

In Landscape 1:

_____

In Landscape 2:

_____

In Landscape 3:

_____

Match the Silhouettes

An artist was walking around a museum.
Five giant geometric solids were on display.
The artist stopped here and there
to draw silhouettes of what she saw.

She drew one silhouette of each solid
from somewhere on the museum floor.

Match each silhouette to one of the geometric solids.

Geometric Solids

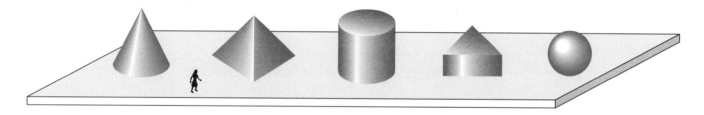

Letter: _____ _____ _____ _____ _____

Silhouettes

A   B   C   D   E

Name _____

Drawing Silhouettes: An Introduction

Use cubes to make this building.

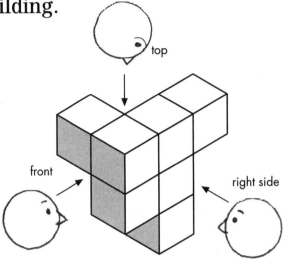

Two silhouettes of the building are shown below.
One was seen from the front, and one from the right side.
The silhouettes are drawn on graph paper so we can see
where the cubes are.

What do you think the top silhouette looks like?

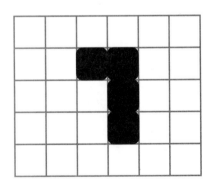

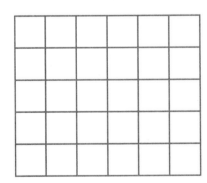

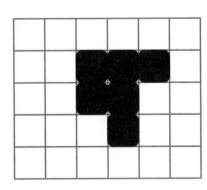

Front   Top   Right side

Front, Top, and Side Silhouettes

Make each building with cubes.
Then draw the silhouettes for both.

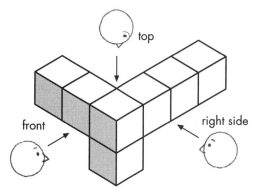

1.

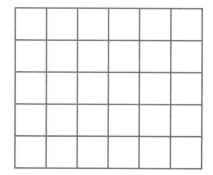

Front

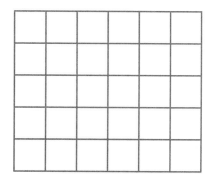

Top, as seen from the front

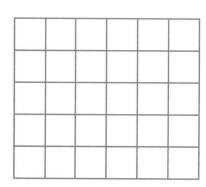

Right side

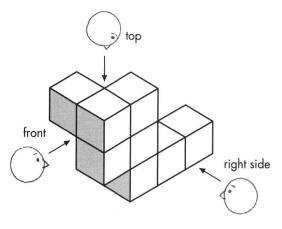

2.

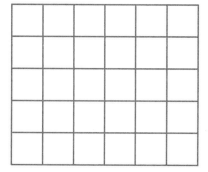

Front

Top, as seen from the front

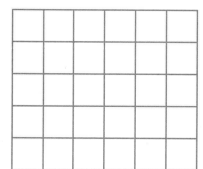

Right side

© Dale Seymour Publications®

Investigation 2 • Sessions 3–4
Seeing Solids and Silhouettes

Drawing Silhouettes A and B

A. Try to draw the three silhouettes for this building. Don't use cubes.

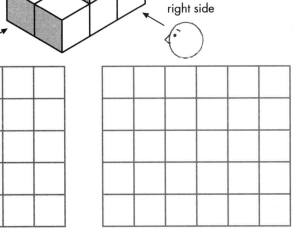

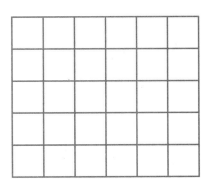

Front	Top, as seen from the front	Right side

- -

Name _____ Date _____

B. Make the building with cubes. Then draw the three silhouettes.

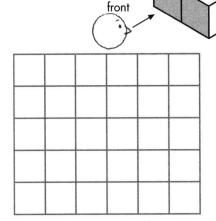

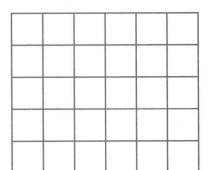

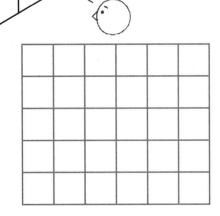

Front	Top, as seen from the front	Right side

Drawing Silhouettes C and D

C. Try to draw the three silhouettes for this building. Don't use cubes.

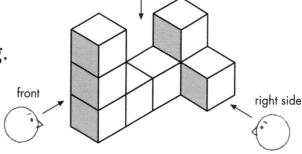

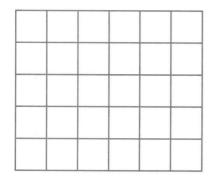

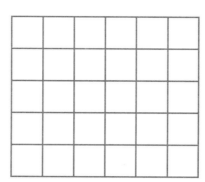

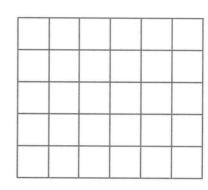

Front   Top, as seen from the front   Right side

- -

D. Make the building with cubes. Then draw the three silhouettes.

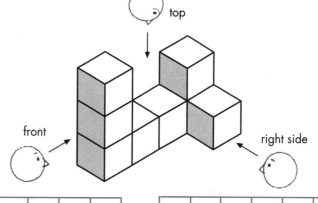

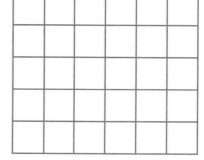

Front   Top, as seen from the front   Right side

Investigation 2 • Sessions 3–4
Seeing Solids and Silhouettes

Puzzles: Building from Silhouettes

For each puzzle below, construct a cube building that makes the three silhouettes. Do any other buildings also make these silhouettes?

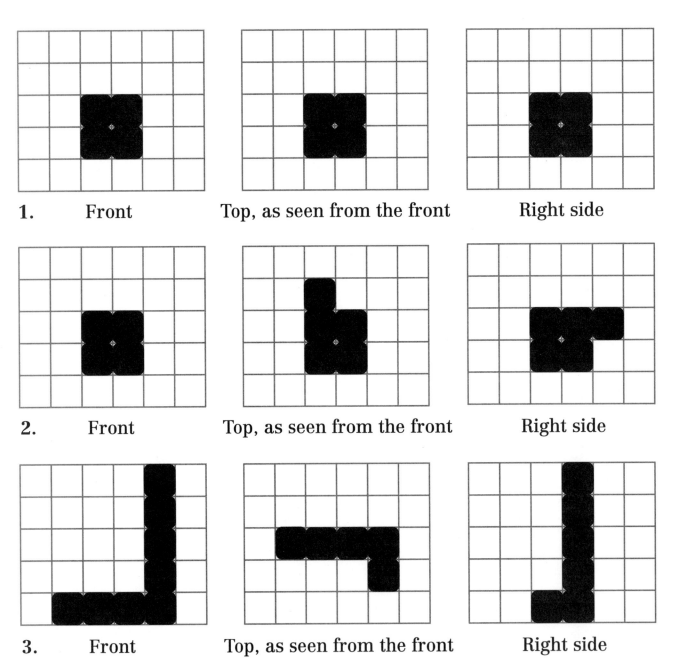

1.   Front   Top, as seen from the front   Right side

2.   Front   Top, as seen from the front   Right side

3.   Front   Top, as seen from the front   Right side

Challenge: How many different cube buildings make the three silhouettes in puzzle 1? in puzzle 2? in puzzle 3?

84

Cube Buildings

Use Student Sheet 16. Draw the silhouettes of these three buildings from the front, top, and right side.

1.

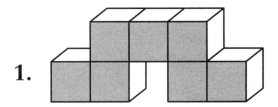

2.

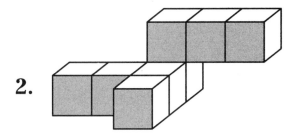

3.

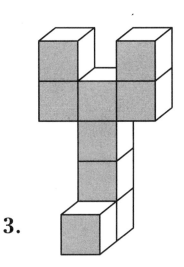

Cube Silhouettes

Draw front, top, and right-side silhouettes for the three cube buildings. Put the number of the building above its silhouettes.

Building _____

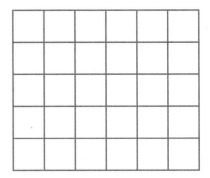 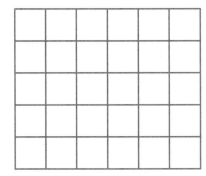 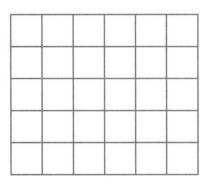

Front   Top, as seen from the front   Right side

Building _____

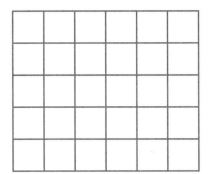 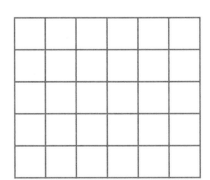 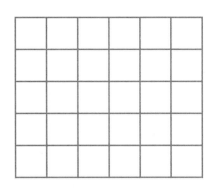

Front   Top, as seen from the front   Right side

Building _____

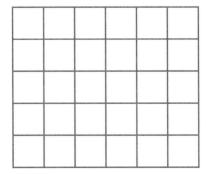 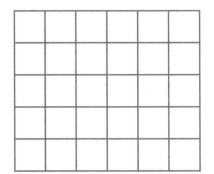 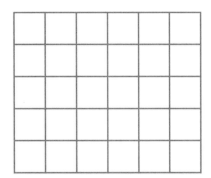

Front   Top, as seen from the front   Right side

Mystery Silhouettes

List objects you find at home that make silhouettes
that are these shapes. The size of the silhouette
doesn't have to match.

1.    Object(s) I found that make a square
silhouette:

_____

_____

2.    Object(s) I found that make a triangular
silhouette:

_____

_____

3. 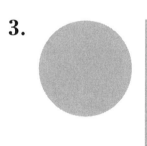   Object(s) I found that make both a rectangular silhouette and a circular silhouette:

_____

_____

On a separate sheet of paper or on the back of this
sheet, draw a silhouette of an object you find at home.
The class will try to guess the object you chose from
the silhouette you draw.

87

Different Views of a City

This map shows the top view of a cube city.

The eight buildings shown are made from interlocking cubes.

The number on each building tells how many cubes high that building is.

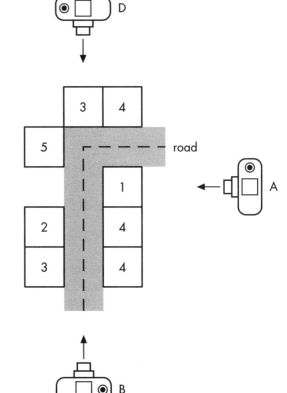

A photographer flew around the city in a helicopter and took four silhouette photographs.

The photographs were taken from points A, B, C, and D (looking in the directions of the arrows).

The resulting silhouettes are shown here. Below each one, write the letter of the point where it was taken.

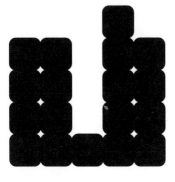   

1. ____   2. ____   3. ____   4. ____

More Cube Buildings

Use Student Sheet 16. Draw the silhouettes of these three buildings from the front, top, and right side.

1.

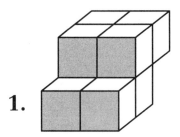

2.

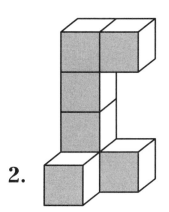

3.

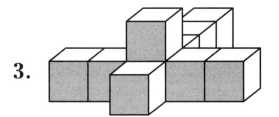

Building Instructions Set A:
3-D Picture

Make a building out of ten cubes by looking at its picture below.

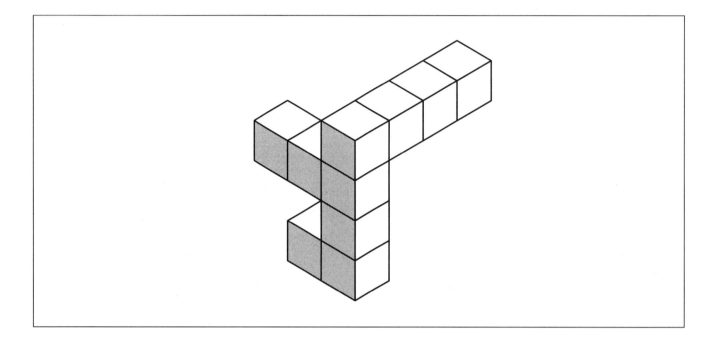

Are these good instructions for making a cube building?

_____

_____

Tell what is good and what is bad about these instructions.

_____

_____

_____

_____

_____

Building Instructions Set B:
Three Straight-On Views

Make a building out of ten cubes by looking at
the three pictures of it below.

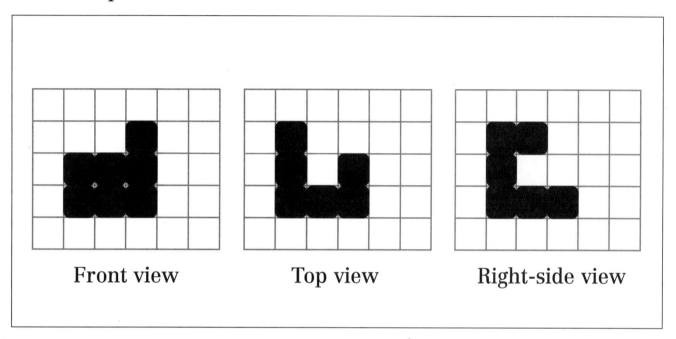

| Front view | Top view | Right-side view |

Are these good instructions for making a cube building?

_____

_____

Tell what is good and what is bad about these instructions.

_____

_____

_____

_____

_____

_____

Building Instructions Set C:
Two Straight-On Views

Make a building out of ten cubes by looking at
the two pictures of it below.

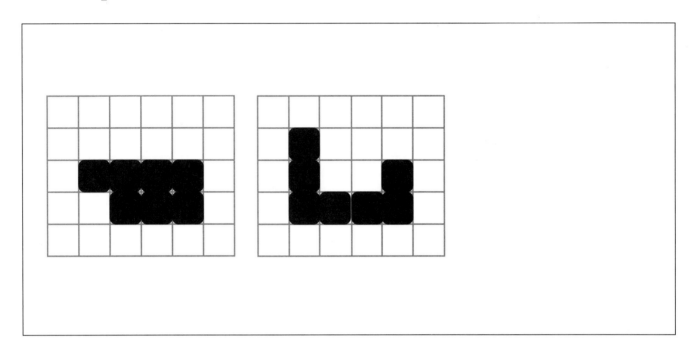

Are these good instructions for making a cube building?

_____

_____

Tell what is good and what is bad about these instructions.

_____

_____

_____

_____

_____

Building Instructions Set D: Layer-by-Layer Plans

Make a building out of ten cubes by looking at the plans below.

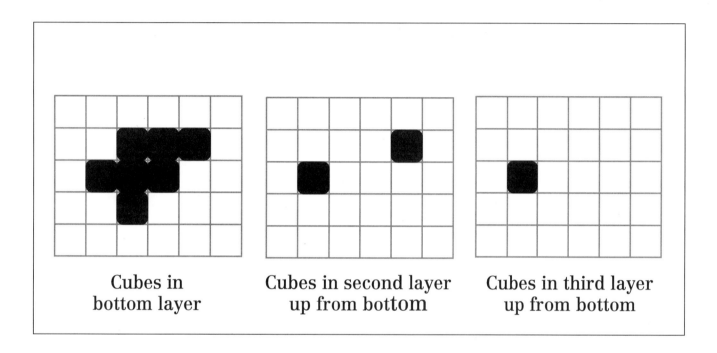

| Cubes in bottom layer | Cubes in second layer up from bottom | Cubes in third layer up from bottom |

Are these good instructions for making a cube building?

_____

_____

Tell what is good and what is bad about these instructions.

_____

_____

_____

_____

_____

Building Instructions Set E: Written Directions

Make a building out of ten cubes by
reading the written directions below.

Take five cubes and attach them to make a
straight line. Take another three cubes and attach
them to make a straight line. Lay both lines flat on
a table, with both horizontal. Place the shorter
line of cubes behind the longer line so that the
shorter is placed evenly between the ends of the
longer. Attach these two lines together.

Now connect the two remaining cubes. Hold
these two cubes so that they are sticking up from
the table—they should be vertical. Now attach
these two cubes to the top of the last cube on the
right of the shorter line of cubes.

Are these good instructions for making a cube building?

_____

_____

Tell what is good and what is bad about these instructions.

_____

_____

_____

_____

_____

Quick Image Geometric Designs

1. Cut out the Quick Images below.

2. One person picks a shape and turns it face up for 3 seconds.

3. The other person tries to draw the shape from the image in his or her mind.

4. Repeat Steps 1 and 2 with the same shape so that the second person can revise the drawing.

5. Reveal the shape, and compare it with the drawing. How did the second person see the image each time it was shown? How did he or she remember what the shape looked like?

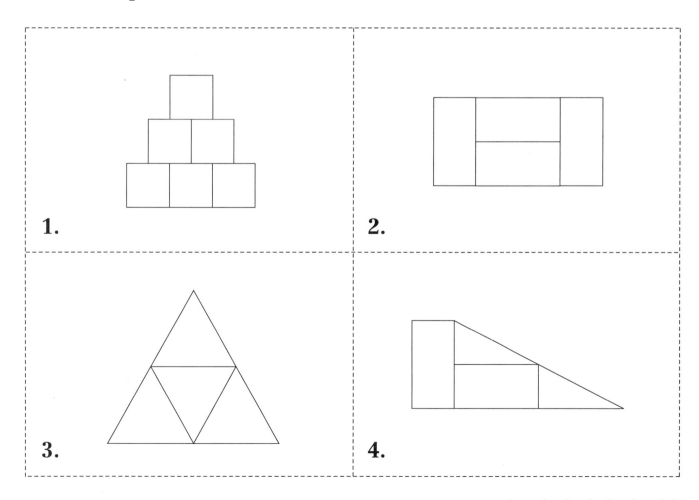

1.

2.

3.

4.

Quick Image Dot Patterns

1. Cut out the Quick Images below.

2. One person picks a pattern and turns it face up for 3 seconds.

3. The other person tries to draw the pattern from the image in his or her mind.

4. Repeat Steps 1 and 2 with the same pattern so that the second person can revise the drawing.

5. Reveal the pattern, and compare it with the drawing. How did the second person see the image each time it was shown? How did he or she remember what the pattern looked like?

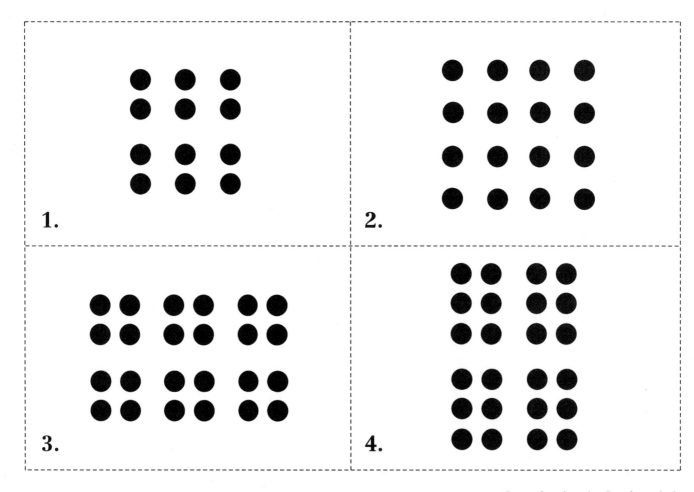

Make a Quick Image

On a sheet of graph paper, color 20 black squares in an arrangement to use for Quick Images. Use the grid-lines to help line up your squares in an arrangement you think the class will remember. Try out your Quick Image with a family member to see how difficult it is for another person to remember and copy.

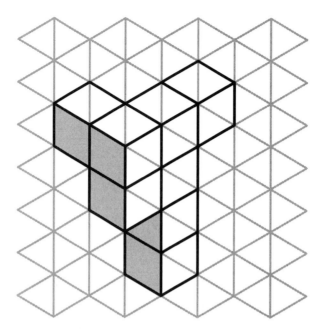

ONE-CENTIMETER GRAPH PAPER

100

THREE-QUARTER-INCH GRAPH PAPER

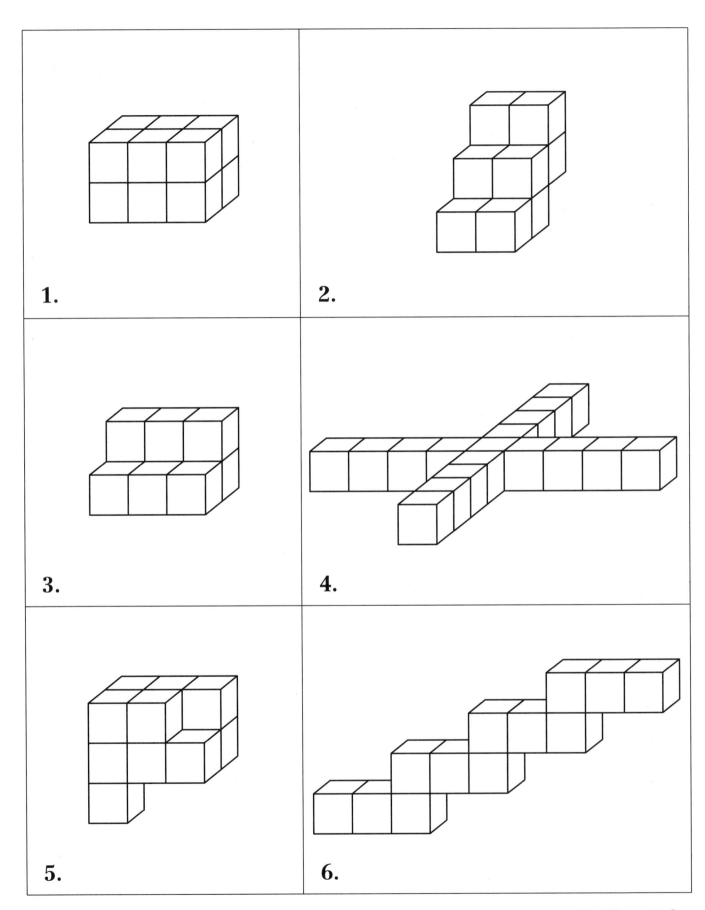

1.

2.

3.

4.

5.

6.

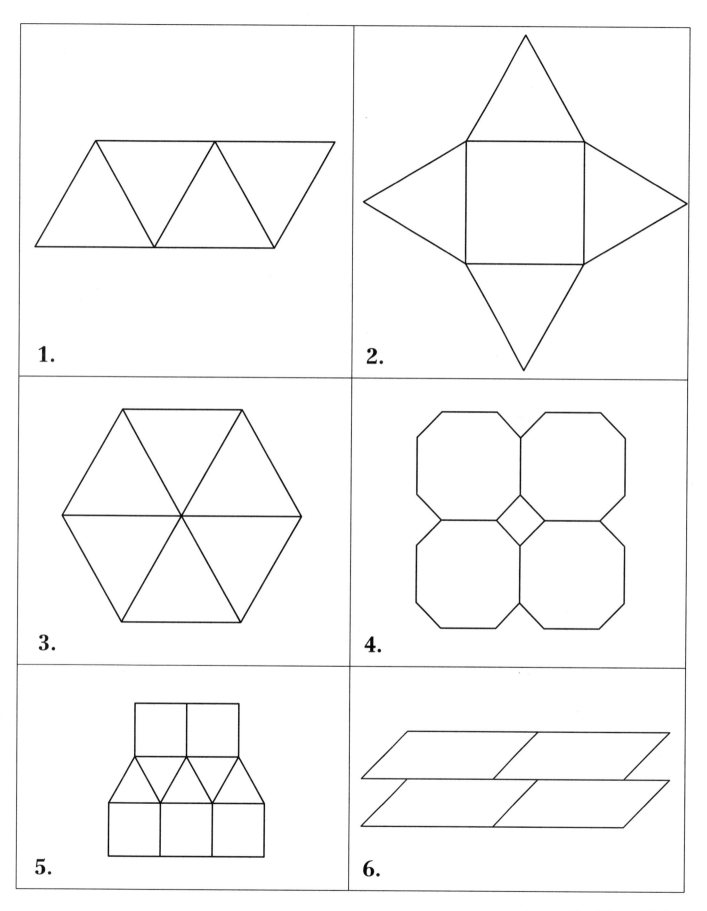

1.

2.

3.

4.

5.

6.

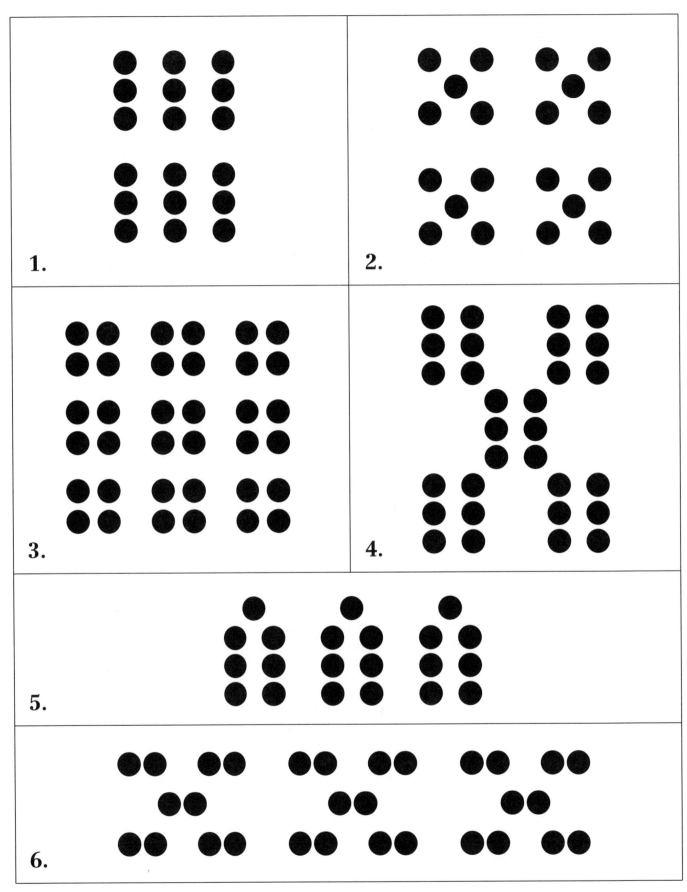

Practice Pages

This optional section provides homework ideas for teachers who want or need to give more homework than is assigned to accompany the activities in this unit. The problems included here provide additional practice in learning about number relationships and in solving computation and number problems. For number units, you may want to use some of these if your students need more work in these areas or if you want to assign daily homework. For other units, you can use these problems so that students can continue to work on developing number and computation sense while they are focusing on other mathematical content in class. We recommend that you introduce activities in class before assigning related problems for homework.

Close to 100   This game is introduced in the unit *Mathematical Thinking at Grade 4*. If your students are familiar with the game, you can simply send home the directions, score sheet, and Numeral Cards so that students can play at home. If your students have not played the game before, introduce it in class and have students play once or twice before sending it home. For more challenge, students can try the variation listed at the bottom of the sheet. You might have students do this activity two times for homework in this unit.

Ways to Count Money   This type of problem is introduced in the unit *Mathematical Thinking at Grade 4*. Here, three problem sheets are provided. You can also make up other problems in this format, using numbers that are appropriate for your students. Students find two ways to solve each problem. They record their solution strategies.

Story Problems   Story problems at various levels of difficulty are used throughout the *Investigations* curriculum. The three story problem sheets provided here help students review and maintain skills that have already been taught. You can also make up other problems in this format, using numbers and contexts that are appropriate for your students. Students solve the problems and then record their strategies.

How to Play Close to 100

Materials

■ One deck of Numeral Cards

■ Close to 100 Score Sheet for each player

Players: 1, 2, or 3

How to Play

1. Deal out six Numeral Cards to each player.

2. Use any four of your cards to make two numbers. For example, a 6 and a 5 could make either 56 or 65. Wild Cards can be used as any numeral. Try to make numbers that, when added, give you a total that is close to 100.

3. Write these two numbers and their total on the Close to 100 Score Sheet. For example: 42 + 56 = 98.

4. Find your score. Your score is the difference between your total and 100. For example, if your total is 98, your score is 2. If your total is 105, your score is 5.

5. Put the cards you used in a discard pile. Keep the two cards you didn't use for the next round.

6. For the next round, deal four new cards to each player. Make more numbers that come close to 100. When you run out of cards, mix up the discard pile and use those cards again.

7. Five rounds make one game. Total your scores for the five rounds. LOWEST score wins!

Scoring Variation

Write the score with minus and plus signs to show the direction of your total away from 100. For example: If your total is 98, your score is –2. If your total is 105, your score is +5. The total of these two scores would be +3. Your goal is to get a total score for five rounds that is close to 0.

Close to 100 Score Sheet

Name _____

Game 1   Score

Round 1:   ___ ___ ___ + ___ ___ ___ = _____   _____

Round 2:   ___ ___ ___ + ___ ___ ___ = _____   _____

Round 3:   ___ ___ ___ + ___ ___ ___ = _____   _____

Round 4:   ___ ___ ___ + ___ ___ ___ = _____   _____

Round 5:   ___ ___ ___ + ___ ___ ___ = _____   _____

TOTAL SCORE   _____

Name _____

Game 2   Score

Round 1:   ___ ___ ___ + ___ ___ ___ = _____   _____

Round 2:   ___ ___ ___ + ___ ___ ___ = _____   _____

Round 3:   ___ ___ ___ + ___ ___ ___ = _____   _____

Round 4:   ___ ___ ___ + ___ ___ ___ = _____   _____

Round 5:   ___ ___ ___ + ___ ___ ___ = _____   _____

TOTAL SCORE   _____

0	0	1	1
0	0	1	1
2	2	3	3
2	2	3	3

4	4	5	5
4	4	5	5
<u>6</u>	<u>6</u>	7	7
<u>6</u>	<u>6</u>	7	7

8	8	9	9
8	8	9	9
WILD CARD	WILD CARD		
WILD CARD	WILD CARD		

Practice Page
Seeing Solids and Silhouettes

Practice Page A

Find the total amount of money in two different ways.

> 1 half dollar
> 9 nickels
> 6 pennies
> 4 dimes

Here is the first way I found the total amount of money:

Here is the second way I found the total amount of money:

Practice Page B

Find the total amount of money in two different ways.

        4 quarters
        8 pennies
        9 nickels
        2 dimes

Here is the first way I found the total amount of money:

Here is the second way I found the total amount of money:

Practice Page C

Find the total amount of money in two different ways.

> 1 half dollar
> 5 quarters
> 4 dimes
> 3 nickels

Here is the first way I found the total amount of money:

Here is the second way I found the total amount of money:

Practice Page D

For each problem, show how you found your solution.

1. Five people want to share 32 pencils equally. How many pencils does each person get?

2. I have 32 pencils. I want to put them in boxes that hold 5 pencils each. How many boxes will I need?

3. I bought 5 boxes of pencils. I spent 32 dollars. How much did I spend on each box of pencils?

Practice Page E

For each problem, show how you found your solution.

1. Pablo uses color folders to organize his papers.
   He has 13 folders in each of four different colors.
   How many folders does he have altogether?

2. There are four books in a new science fiction series.
   In our class, 13 students each own the whole series
   of books. How many books do they own altogether?

3. Stephanie bought four "baker's dozen" rolls.
   A baker's dozen is 13 rolls. How many rolls
   did Stephanie buy?

Practice Page F

For each problem, show how you found your solution.

1. Twenty-eight gymnasts are going on a tour.
   They will perform together and individually.
   Each gymnast wants two minutes alone. How
   much "together time" will they have if their
   entire show is ninety minutes long?

2. The gymnasts get very thirsty during their
   performances. How many bottles of water are
   needed if they each drink three bottles during
   the show?

3. The gymnasts use ninety towels during each
   performance. How many towels does each
   of the twenty-eight gymnasts use?